STAPLE IT!

STAPLE IT!
Easy Do-It Decorating Guide

by Iris Ihde Frey

CROWN PUBLISHERS, INC., NEW YORK

Library of Congress Cataloging in Publication Data

Frey, Iris Ihde.
    Staple It!

    Includes index.
    1. House furnishings.   2. Interior decoration—
Amateurs' manuals.   3. Staples and stapling
machines.   I.   Title.
TX315.F73   1979   645   74778-11588
ISBN 0-517-532549 (cloth)
ISBN 0-517-532557 (paper)

Designed by Rhea Braunstein

10 9 8 7 6 5 4 3 2

Contents

Foreword

A gold watch is not the classic farewell gift to a display person leaving a job. Sometimes it is an elaborately trimmed, rhinestone-studded staple gun—with good reason. As the baton is to the conductor or the typewriter is to the author, the staple gun is to the display person. The staple gun and stapler can replace hammer and nails, needle and thread, sewing machine— and do some projects that no other tool can.

Who hasn't considered the inviting room setting in a store window, the colorful center-spread in a decorating magazine, or the smashing home in a television show and thought, "I'd like to move right in." But chances are the room in the magazine was constructed in a photographer's studio, the television set was struck minutes after the show was taped, and, of course, the room setting in the store window had neither central heating nor hot and cold running water.

The people responsible for creating these "environments on location" cope with very real problems. They have budget limits, time pressure, and the never ending treadmill of tomorrow's show, next month's issue, and the seasonal change of room settings. They work in limited space, with the barest of cubicles lacking in architectural detail, yet they regularly produce crowd-stopping rooms. Of necessity, these professionals have developed a wealth of techniques for creating instant interior design. Most of these techniques involve one mighty tool—the staple gun.

It is no secret that a great many of the dazzling effects in the decorative arts are achieved by illusion. Since the days when the laws of perspective were first developed, artists have played with the effects of illusion, from trompe l'oeil (deceive the eye) paintings to the dazzling effects of more recent popular op art. This book is not just about creating an illusion, nor is it about Band-Aid decorating. It is about practical, common-sense, time- and money-saving techniques that I learned or developed while working in Bloomingdale's and other department stores, and in television studios, including NBC.

There is no project in this book that requires machine sewing. This is not to suggest that sewing is a thing of the past, a less preferable technique, nor that the book was written by a nonsewer. There are already many books that deal with the subject of sewing for the home—this book shares the tricks of the trade that eliminate the need for it. There are instances where machine sewing is an alternative method, but a sewer will know where this is the case and can make a choice.

The intent of this book—to make home decorating possible for everyone—and the current trend in home furnishings are very much to the point. The new look has been described as wrapped, amorphous, and relaxed. Where yesterday's chairs were fitted, seamed, and coaxed into a glovelike fit we now celebrate gathers, tucks, and draping.

A look through an issue of an upholstery trade magazine turns up pages of advertisements for such items as "instant upholstery kit—attaches welting, gimp, skirts, and face panels in seconds with hot melt adhesive, plus electric rapid stapling gun." This might surprise the skeptic who thinks that the use of staple guns requires an apology. A visit to a furniture manufacturer would show that stapling is an accepted construction method.

Indeed, many houses today are built with heavy-duty and electric staplers designed for the construction industry. This book assumes that the reader already has a roof overhead and won't go into that. Included is a section, however, that describes a simple method for paneling a room with wood using a special kind of stapler that shoots nails.

Each section has photographs of rooms that are examples of excellence of contemporary design. The elements in these rooms have not necessarily been done with a staple gun. Some of them were executed using traditional methods of European and continental craftsmanship. But in each case, the photograph was selected as the inspiration of a beautiful environment that the reader could personally execute.

As the book was being written it clearly divided itself into natural divisions—things to sit on, to lie on; things we set things on; walls; ceilings, and so on. Each section has step-by-step photos and diagrams of projects. Some of the projects are so simple that they require little direction. Some of them require some tools or supplies and a fair amount of time. Some projects even require a good deal of patience. Some deal with before-and-after rejuvenations; others are for new creations. But all have been executed from start to finish by the author. The photographs are proof that they are not pie-in-the-sky ideas for impossible projects but very real ones that can be duplicated by the reader.

Doing it yourself allows you to make all the choices—size, color, and style. The economy of it is already well known. I hope the reader will gain from this book the greatest reward—that unmatched good feeling of accomplishment that comes from using your own creative hands.

1 All about Staplers

What's So Great about Staplers?

The stapler is like a carpenter, with his nails and hammer, turned into a computerized robot. The staple is the equivalent of two nails driven simultaneously, rapidly, and with total accuracy. An advantage in using staples over nails is that the thin width of the staple puts a much smaller hole in the material being fastened. A nail as thin as a staple would be impossible to drive, as it would bend before it could be hammered into place. Not only does the U shape of the staple provide double holding points, but the crosspiece serves as an additional support in holding and in preventing tearing. Unlike nailing, a beginner can place a staple with dead accuracy—it will enter exactly where it is directed. Another big advantage is that the stapler is designed to work on the principle of stored energy, provided by a powerful spring, making it an almost effortless operation. The staple gun is a great leveler of mankind. Because its penetrating power is self-contained, a ninety-eight-pound weakling gets as much power from the stapler as a muscular weight lifter. Of course, the heavy-duty staple gun requires considerably more effort and a large enough grip to operate one. Another plus is that stapling is a one-hand operation. The other hand is free to position, stretch, and control. And finally, it is very fast. The staple is anchored with one movement; hammer and nails need repeated blows.

The advantages of using staples have long been recognized. More than one hundred years ago inventors were designing "Devices for Driving Staples." In 1872 James Tomlinson was awarded a patent for his cast-iron tool of service for many uses but especially of value in driving staples into barrel-hoops. A big drawback of Mr. Tomlinson's device was that only one staple at a time could be loaded. Another disadvantage was that it wasn't automatic. After the staple was loaded, the device was held steady while a hammer whomped a plunger until the staple was in place.

Opposite page: Using one vibrantly striped fabric, Chicage interior designer J. Neil Stevens turned a tiny hallway into a cozy library and game area. Fabric covers table, walls, ceiling, chairs, and bookshelves.
Courtesy Celanese House

How Staplers Work

Inside the staple gun are several springs. The powerful hammer spring is located vertically at the front of the gun. The hammer spring drives a ram that shoots out the staple. This is the way it works: as the lever handle is pressed down, the hammer spring is compressed. When the lever is down all the way, a mechanism releases the spring. This causes the ram to drive the staple out. The lever returns to its up position when the hand releases it. As soon as a staple has been shot out, the row of staples (held under tension by a smaller horizontal spring) moves forward so that the next staple is under the ram ready to go.

Because loose staples would be difficult to handle, stacks of staples are lightly glued to hold together during loading. This coating is not heavy enough to interfere with the operation of the ram. The following illustrations demonstrate what happens inside the staple gun.

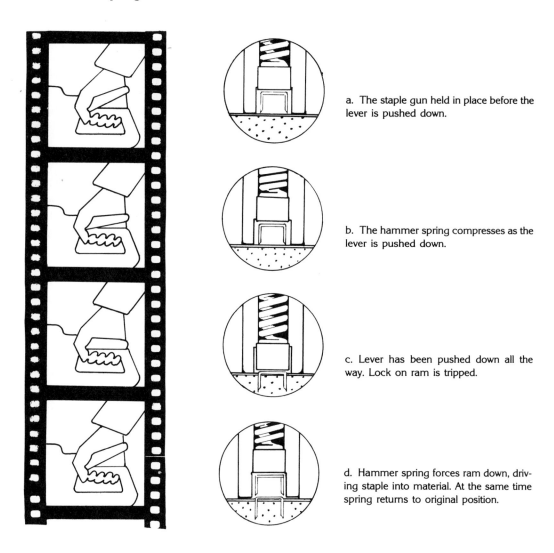

a. The staple gun held in place before the lever is pushed down.

b. The hammer spring compresses as the lever is pushed down.

c. Lever has been pushed down all the way. Lock on ram is tripped.

d. Hammer spring forces ram down, driving staple into material. At the same time spring returns to original position.

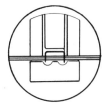 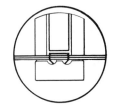 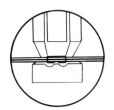

The desk stapler operates the same as the staple gun except that an anvil underneath forces the legs of the staple to bend back up.

The operation of the staple gun and the desk stapler is the same, but there is a difference in what happens to the staple after it leaves the mechanism. The bottom half of the desk stapler has a metal plate called an anvil. The anvil has two depressions corresponding to the prongs of the staples. When the ram forces the staple through the material, the prongs of the staple legs hit the depressions in the anvil and are bent in and up toward the ram. This causes the staple to clamp the material tightly. Some business-quality staplers have anvils that crimp the staple in patterns or make it fold outward rather than in. Some of these commercial staplers are very powerful and can staple through many layers of cardboard or through a hundred sheets of paper at a time.

Most desk staplers have anvils that can be removed or swung out of the way. This means they can be used as staple guns. While they are more awkward to use than staple guns, this adaptability often comes in handy. They use smaller and shorter staples and at times this can be an advantage.

Kinds of Staple Guns

All of the many makes of staple guns fall into either the heavy-duty or light-duty class. Heavy-duty guns are larger and heavier, and take staples of a thicker gauge of wire—.05 inch thick. A heavy-duty staple gun takes a larger range of staple lengths from ¼ inch to 9/16 inch. Because a heavy-duty gun has a stronger spring it has increased penetration power. It also requires more strength to operate, and some models require a large-size hand in order to grip it comfortably. Because of their increased built-in power, the heavy-duty guns tend to recoil at times, requiring one hand to operate and the other to apply downward pressure.

A light-duty staple gun, which weighs less and is smaller in size, takes staples .03 inch thick. Staple lengths for light-duty guns range from 3/16 inch to ½ inch. The average number of staples held by a light-duty gun is 120 while a heavy-duty gun, because of the thicker staples it takes, can hold only 85.

With few exceptions, staples fit only the stapler for which they are manufactured. The wrong-size staple can cause jamming and might actually damage the tool. Even if the staple of another brand can be found it would be a risk to use, as all guarantees specifically state that repairs or replacements will not be made if the damage is caused by using the wrong staples.

Special Attractions and Attachments

A variety of special features has been incorporated into the design of staple guns, and separate attachments have been developed to assist in specific jobs.

Perhaps most common is a locking device that prevents firing. This is usually built in. Another feature that comes both attached and separate is a staple remover. A very effective staple remover that operates with jawlike action is sold in stationery stores. Some guns have a gauge that shows the number of staples left in the gun. One manufacturer's helpful feature is a window showing not only the amount of staples left

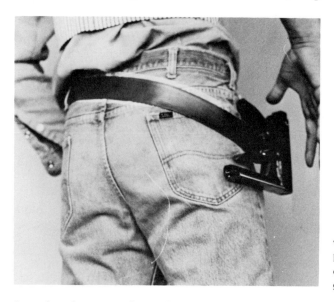

With gun slung from holster belt, this fellow is ready for action. An interesting variety of accessories is available for staple guns.

but also their size. Some heavy-duty models have a power adjustment that gives extra force to the staple penetration. Several manufacturers sell a screen attachment—a metal piece that clips onto the bottom front of the gun. Its teeth grip the mesh of the screening and allow it to be tightly stretched. One manufacturer sells a kit for stretching artist's canvas. In this case the teeth on the attachment grip the material and pull it taut and wrinklefree while stapling. This attachment is also recommended for mounting needlepoint and embroidery. Also available is a window shade attachment—a curved steel plate that steadies the gun while stapling into dowels or shade rollers.

Electric staple guns have been used for years by the trade for upholstering and carpentry. The introduction of lightweight, inexpensive models now makes the electric staple gun a likely tool for the home workshop. The electric staple gun does the job of a heavy-duty stapler by simply pushing a button. It works on heavy plaster walls and hardwood, and takes a range of staple lengths. The electric staple gun is a true power tool—it should be used with care. Manufacturers recommend that safety glasses be worn while using it.

Selecting a Staple Gun

This array of choices may be confusing to someone about to buy a first staple gun. What Will Rogers said about men—he never met one he didn't like—I can say

The Purkersdorf Arm Chair, designed in 1903 by Josef Hoffmann. The frame is white lacquered wood, the seat is woven black and white webbing that forms a checkerboard pattern. Three-quarters of a century later it remains a classic example of good design. The webbing is attached in straightforward functional manner—stapled to the bottom of the seat. *Courtesy International Contract Furnishings.*

This electric staple gun makes quick and effortless work of a reupholstery job on a chair. Five different staple sizes make this same tool suitable also for big home maintenance projects such as roof coverings and ceiling-tile installations. *Courtesy The Swingline Company*

about staple guns. The brands I used for the projects in this book included, in alphabetical order, Arrow, Hansen, Sears, and Swingline. Some of them were acquired in the process of writing the book and some had been in the family for years. They all worked well. The attachments are helpful but not essential. Having such a number of tools available turned out to be a luxury because a variety of staple sizes was always loaded and ready.

If all other things were equal, the staple gun to choose would be the most versatile one. This would be the one that took the largest variety of staples and had the most attachments. But the staple guns that best fit this description are the heavy-duty ones—and they have the drawback of being more tiring to use because of their weight and the additional strength required to operate. The most important factor in selecting a staple gun is that it be comfortable to operate. While a big strong hand can adapt itself to any stapler, a small hand may not manage all of the large models on the market. When a consumer product magazine tested staple guns for holding power, their conclusion was that the length of the staple was more important than its gauge (whether it came out of a light-duty or heavy-duty gun). They also noted that some testers found guns that pinched, but this likely has to do with the compatibility of hand size and gun size. Trying out the gun is the only way to judge if it fits. If children are around, get a gun with a lock. Finally, a consideration in choice is that the staples required for it be readily available.

A light-duty gun can handle most jobs and is the easiest to use. No home should be without one. If only one gun had to be chosen to do all the projects in this book, it would have been a light-duty one. After it has proved itself to be an indispensable tool, a heavy-duty gun—either manual or electric or both—will be a practical addition.

Short Course in Using a Staple Gun

Hard to believe, but there are staple guns that load from the top, from the bottom, from the back, and from the front. As each gun has its own distinctive method no attempt will be made here to explain loading and unloading. But the time spent studying the manufacturer's directions will be very worthwhile. Although the operation may appear to be self-evident, special features or shortcuts may be discovered by reading the directions. Learning how to quickly lock and unlock the gun will add to its safe use and subtract from the user's potential frustration.

Aiming

Study the staple exit point (corresponding to the plunger's location) in relation to the front and side edges of the gun. It is important to know exactly how far from the edges—front and side—the staple is set in. If a row of staples is to be placed following an edge, it is fairly easy to line them up evenly. An edge to measure from is not always visible. In this case a pencil or chalk line marked on the material or the base being stapled into can be a great help. Another way to place staples evenly is to put a piece of masking tape on the side of the gun and use it as a marker. With staple gun loaded, unlocked, placement determined, the rest follows naturally and easily. Most often the staple gun is operated by squeezing the lever with the fingers. Stapling into a material of high density requires more effort, and it may be necessary to push down on the gun with the weight of the body behind the palm of the hand. For solid driving hold the staple gun in a flat and snug position against the material to be stapled. If staples are to be driven into hard wood, or if very long staples must be driven, it may be difficult to sink the staples flush with the surface. The gun likely has enough power but may tend to jump back or recoil. If this happens, the full power of the gun is reduced because it is difficult to press it hard against the surface while at the same time pressing the lever. If this is the case, use the gun with two hands. Press one hand down against the top of the gun while the other pulls the lever.

Staple Length

If, after using the method just described, the staples do not go in all the way, it usually is a signal to switch to a shorter staple. On the other hand, check to see that a staple is not too long for the job. On some bases such as the pressed cardboard ones, the shortest staple available may be so long that it will puncture through to the surface of the material. In this case either increase the thickness of the base with an additional layer of cardboard or other padding, or switch to a desk-type stapler with its shorter-length staple. Staple length must also be considered for its suitability for the job to be done. At times the staple is required simply to hold material in place until the glue dries. For some decorative uses the staples secure the material but have no stress placed on them. If the staples have to stand up to wear and tear—as on the upholstery of a seat—they need to be long enough to provide adequate holding power.

Jamming

After depressing the trigger, let it come up all the way; not doing so is an invitation to jamming. If the handle is not released all the way, the next staple may advance too soon. Jamming tends to happen when stapling at great speed. It also happens if staples of different lengths get loaded into the channel at the same time. If jamming happens, first try shooting the gun several times. Don't force the mechanism. Next check to see if the gun is really jammed or is just out of staples. Remove all staples from the channel, fire the gun several times, and reload. If after these steps the problem hasn't cleared up, the chances are that a staple is stuck in the mechanism. NEVER SQUEEZE THE TRIGGER ON A STAPLE GUN WHILE IT IS AIMED AT FACE OR EYES. If a stuck staple is the problem, a plier may be needed to remove it. Squeeze trigger a few times after removing staple but before reloading. More rarely the jamming could be caused by improperly shaped or bent strips of staples. If they are out of shape, they can cause a jam by gripping the channel and not advancing. Not to worry—although dozens and dozens of boxes of staples were consumed by staplers in the preparation of this book, not one case of jamming occurred.

Baste Stapling

Just as upholsterers do what they call baste tacking—hammering a tack lightly and temporarily while working the fabric into place—staple gunners can do what is called baste stapling. With a little practice the staples can be controlled so that they penetrate as little or as much as is needed. Partially inserted baste staples can be easily removed with a screwdriver, staple remover, or fingertips. It is a technique that should be routinely used. Care in positioning is often the factor that determines success or failure. Instead of holding the staple gun flat against the material being stapled, tilt it so that only an edge—either the front or one of the sides—rests on the material. This prevents the staple from shooting with full force, so it does not enter completely.

On the other hand, to get increased holding power, shoot a second staple on top of the first one so that a + is formed.

With so many "Dos" in the preceding section, it seems a good idea to add at least one "Don't."

Don't staple into a water bed.

"If it takes forever . . . I will Wait for You . . ." is the title of an unusual installation in a Berlin art gallery. The artist, Colette, uses parachute silk and a staple gun instead of paints and a brush. *Courtesy Eugenia Cucalon Gallery*

2 Materials and Techniques

Tools

Besides the staple gun the tools used in this book are few. Just as a fine meal can be prepared without an elaborately equipped kitchen, the projects in this book can be successfully duplicated with very few simple tools. Power equipment—electric drills, saws, and so on—is helpful but not necessary. Lumberyards will make the cuts needed at the time lumber is purchased. They charge by the cut. As lumberyard personnel appear always to be rushed, it is wise to shop at other than peak busy times or just before closing, and to have written down exactly what lengths and pieces are required as well as a sketch of the project. These are the tools most often used:

Utility knife and blades (mat knife) Staple remover
Saw Single-edge razor blades
Scissors Heavy-gauge pins
Steel square or T square Upholstery needle
Measuring devices

Supplies

Coverings

Many different types of coverings are used for the projects in this book. They range from the sheer chiffon used on a paneled screen to the woven straw headboard. Felt, which is not woven but made of compressed bits of wool, is good to work with because it can't ravel and needs no turning under. Knits are often a good choice because their give and stretch easily conform to covering curves and bulges. Textiles are divided into two classes—decorator, or home furnishings fabrics, and fashion fabrics. The differences between the two should be understood, but then the class lines can be disregarded and the fabrics used interchangeably provided weight and durability are suitable. One difference between the two is their width. Fashion fabrics are usually 45 inches wide except for woolens, which are 54 to 60 inches. (A few pure cottons are 36 inches wide.) Decorator fabric widths range from 48 to 60 inches. Some drapery and curtain fabric is woven up to 118 inches wide. Sheets and sheeting by the yard are available in bed-size widths. Another significant difference between the two kinds of fabric is that decorator fabrics are planned to have a pattern match at the selvage edge, while fashion fabrics do not necessarily. The section on putting fabric on walls has more information on pattern repeats. Some fabrics have an indistinguishable right and wrong side and work well when something is to be seen from both sides, as a screen or shutter panel.

Selvages

Use them whenever possible. They can serve as a guideline for insuring straight stapling. If they are not woven so tightly that they pucker, they can serve as the finished sides of window shades or draperies. Cut off, the selvage can be used as a finishing tape to glue over staples. Some decorative fabrics have selvage legends. These are printed notations that may include the date, designer's name, name of the fabric, and a numbered row of the separate colors used to print the pattern. Obviously these selvages cannot serve a decorative purpose and must be turned under or trimmed off.

Grain

The pattern printed on the cloth is not always perfectly lined up with the grain (horizontal and vertical weave) of the cloth. Sometimes it is necessary to make a choice between following the grain or following the pattern. Follow the pattern. A crooked pattern can create a disorienting if not downright dizzying sensation. The grain line on solid colors should not be ignored, especially if it has a prominent texture. If the fabric is not stapled straight "snaking" will occur. This zigzagging will be unattractive, whether it is on a fabric with a prominent grain such as slubbed linen or on a linear pattern such as stripes or plaids.

Plastics

Plastics come in the same wide widths as decorator fabrics. While most of today's plastics have a good recovery rate (they return to their original flat state after being stretched or folded), if they are to be stored for any length of time, roll rather than fold them. If wrinkles persist, laying them flat in the sun or "warming" them with an iron set on a low temperature helps. Plastics lend themselves well to stapling. Although most have a cloth backing, which minimizes ripping, a hole put into plastic means a permanent hole. Because of the small gauge of staples, less puncturing and tearing takes place than if tacks or nails were used. Don't use baste staples on plastic except on edges that will later be covered.

Bases

Following is a list of all of the various bases used for the projects in the book.

FOAM

Foam products are divided into soft and semirigid. Soft foam is what is commonly called foam rubber (real foam rubber is almost extinct). Semirigid foam is the type of material used for ice chests and coffee-to-go cups.

Semirigid foam products are suitable for many projects not subject to hard wear. They are lightweight and relatively inexpensive. They are easy to work with because they can be cut with a utility knife or a single-edge razor blade. The disadvantage is that they dent and edges can crumble.

Best tool in the home for cutting foam rubber is an electric carving knife.

Foam board Some of the trade names for this are Fome-cor and Gatorfoam. Sold in art supply stores in various small sizes but available in sheets up to 4 × 8 feet and in two thicknesses from ¼ inch to ¾ inch.

Insulation panels Sold in lumberyards in various sizes and thicknesses. The larger panels, which are more dense, are light blue or gray and come in 2×8-foot panels or in 3×8-foot panels. These are usually one inch thick. Smaller panels in white of a less dense foam come in 2×4-foot panels in various thicknesses up to one inch. They can be used as the base for wall art, super graphics, or used to panel sections of walls.

Soft foam   Solid pieces of soft foam are used in this book for many of the projects. It comes in densities ranging from firm to soft. Sometimes seats are built in layers of increasing softness. It is available in all thicknesses. Foam sheeting is available in all thicknesses and in all sizes up to room size intended for use as rug pads. Available through mail order catalogues and foam factories.

CARDBOARD

Composition board   Laminated layers of cardboard (actually recycled wood fibers) also known as pulp board and tri-wall cardboard. Sold in lumberyards in 4×8-foot panels. Upson board is one brand that is a mainstay of every department store's display material. It is lightweight but durable. It can be cut with a utility knife. An electric saw cuts it easily but makes a more ragged edge. It comes in ¼-inch and ½-inch thicknesses. Homosote is a similar product, which comes in ½-inch and ⅝-inch thicknesses. In addition to 4×8-foot panels it can be ordered in larger sheets. Because it is composed of bits and pieces, whereas Upson is layered, it tends to crumble more easily at the edges. It can be framed with wood. Either way it is useful for valances, headboards, wall panels, etc.

   Hardboard or chipboard is not used, as it is too brittle and dense to take staples well.

Corrugated cardboard   Free for the collecting, cut from cartons. Large appliance and TV cartons work especially well, as they are made of extra heavy cardboard. One bicycle carton provides enough cardboard for a four-paneled screen. The fluting or ribbed inner construction gives corrugated cardboard its lightweight strength. When more than one layer is used with the ribbing running in opposite directions, surprising strength results.

Tubes   Cardboard tubes ranging in size from a few inches to 12-inch diameter are useful for many purposes. The small ones can be collected for the asking at fabric stores . Large ones can be located at concrete companies, construction suppliers, or carpet stores.

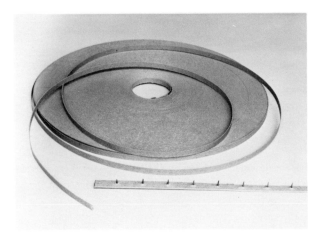

A lifetime supply of upholsterer's tape—a good investment for a few dollars. This tape is not sticky or adhesive in any way. A cardboard strip less than ½ inch wide, upholsterer's tape is used for many projects. To the right below it is another upholstery supply, tack stripping. It is used to enclose the final layer of upholstery.

Upholsterer's tape   Narrow cardboard (less than ½ inch wide) in large rolls, it is used for backtacking. (See page 18.) Available from upholsterers and upholstery supply stores. Substitutes for it are strips cut from corrugated cartons, plasterer's tape, and plastic strapping tape.

PADDING

Polyester batting   Used for both quilting and upholstery purposes. It is available in batts corresponding to blanket sizes for all standard beds. It comes in thicknesses ranging from ½ inch to 1½ inches. Sometimes labeled by 6-, 7- or 8-ounce weights.

Cotton batting   Also used for making quilts. Available in ¼-inch thickness. Heavier cotton batting is available through upholstery suppliers and is about ¾-inch thick.

Foam   Sold in sheeting form or in rolls of various widths and thicknesses. It can be bought in almost any size all the way up to room-size rug pads.

Cotton flannel   Available in fabric stores by the yard, it is often used as an interlining for draperies or to pad the walls of a room being covered in fabric.

WOOD

Plywood   A sandwich of thin layers of wood that have been pressed and glued together. Standard 4×8-foot sheets come in a variety of thicknesses—only ¼ inch and ½ inch were used for the projects here. Although plywood comes in many qualities the cheapest available is suitable for staple gun projects. Roughness or appearance will not matter.

Paneling   Usually grooved on the right side, if the back is real wood (not hardboard) it is an excellent base for projects. It is thinner than plywood—usually about ⅛ inch—and not very expensive. Since the back side will be used and since the project may be covered anyway, damaged panels or seconds, which are often available, will be perfectly suitable. Luan mahogany paneling—usually the cheapest kind of wood paneling sold—even if first grade, is an inexpensive but good base to use.

Unpainted furniture   An increasingly interesting variety of unfinished furniture pieces are available at stores and from mail order houses. They standardly stock wooden cubes in a variety of sizes as well as Parsons tables in all heights and dimensions. Some of the newer styles include floating platform beds and the curved waterfall table.

Crates   Many of yesterday's crates have found a welcome from antique shops. Although the use of wood for packing and crating is almost a thing of the past, there are a few sources left. They might be referred to as the 3 Ms of the wooden crate world—melons, monuments, and motorcycles. Although they look unpromising in their natural states, they hold great decorating possibilities as platforms, tables, stands, and so on.

Lattice   Sanded wood strips 3⁄16-inch thick, they come in a variety of widths.

Furring strips   Narrow strips of lumber attached to the perimeter of a room. Wood lath, sold in bundles, is too rough and unfinished to use for this purpose. Other lumberyard items include dowels and closet poles.

Adhesives

Glue   The usual white glue brands commonly available work well on all the projects that call for white glue. For rugs, use glue sold in floor covering stores.

Spray glue   Aerosol adhesives are sold in art supply and craft shops. For a few projects they are handier than white glue. The spray glue dries very fast and spreads a thin, even layer. Even though they may have an adjustable nozzle the spray tends to get over everything. Spread newspaper on every nearby surface. Be sure to mask out any areas that should not have sticky glue on them. Use only with good ventilation.

Bonding cement   Rubber cement may be used. A stronger cement especially for foam products is available where foam products are sold and through mail order houses.

Fusibles   Modern technology has made these products marvels to use. They stand up to washing or dry cleaning and eliminate a lot of previously necessary sewing. Their use in home furnishing not only is a help for the nonsewer but also eliminates the look of puckered stitching and adds to a crisp tailored look.

Shade lamination material   Especially made for the lamination of decorative fabric to window shades, Stauffer Company's Tri-lam is of room-darkening weight and Tran-Lam is a translucent shade weight. Sold by the yard in 36-, 45-, and 68-inch widths in department stores and shade and fabric shops. Useful for other purposes too, such as lining fabric-covered table tops. The warmth of a dry—not steam—iron releases the adhesive, and as it cools forms a tight bond.

Webs   Trade names are Stitch-Witchery and Poly-Web. A heat-sensitive web of synthetic fibers. Available in strips, sheets, or by the yard in 18-inch widths at sewing counters. The webbing is placed between the materials to be fused, and heat, steam, and pressure are applied. The webbing melts and the two layers surrounding it fuse. A variation of this is the lining material that has heat-sensitive glue on one side only. It adheres but does not melt away.

Iron-on tape   Sold at counters with iron-on patches, tapes are available in colors and can be used as decorative trim.

Masking tape   The beige paperlike tape sold in paint and hardware stores is a frequent helper. It is used to hold glued edges while drying, to serve as the base for self-welting, to cover staples, etc.

Hardware and Notions

Hinges   Used on folding screens and window panels. Bi-fold or universal hinges flex in both directions. To make a very classy bi-fold hinge using an Old World technique updated, turn to page 43.

Brads   Short, fine nails with small heads.

Cup hooks   Small hooks with screw ends. Come in a variety of colors plus silver, pewter, brass. Many uses other than holding cups—shirred fabric, drapery tie backs, dowels. At hardware and five-and-dime stores.

Tack stripping   A sturdy cardboard strip with preinserted tacks, it is an upholstery supply used to attach the final layers on chairs, sofas.

Plumb bob and line   Weight and line uses law of gravity to mark true vertical lines. Hardware stores sell them, but a heavy object such as a hammer tied to a piece of string can serve.

Glides   Metal or plastic buttons that attach to bottom of furniture legs to provide a smooth surface.

Hollow wall fasteners   Plastic expansion anchors, molly screws, toggle bolts listed in order of their holding strength, they are used when hanging objects from walls and ceilings where wooden studs or beams are not available.

Chalk   Playground or blackboard chalk, but not the useless waxed kind called tailor's chalk, is a handy item to have on hand.

Bias tape   Cotton strip cut on the bias and prefolded into a single or double fold. Useful for covering staples, raw edges of fabric panels. Sold at sewing counters.

Hem binding   Woven synthetic tape. Used as bias tape.

Decorative trims   Braid, lace, gimp, welting and double welting, fringes—an endless variety of trims are available. They hide raw edges and staples and can add visual interest, too.

Backtacking

Backtacking is one of the best techniques in the decorator's bag of tricks. Because it is used in so many different ways it deserves a special place in the beginning of the book. Easier to do than to describe, backtacking is also known as blindtacking. Defined, backtacking is a tricky way of layering that hides raw edges and staples.

Pretend that this page is a panel of fabric—which explains the rosebuds. The page on the right is a second fabric panel that has been attached to it. Turn to page 18 to see how it was done.

Fern green and white all over the walls and bed of this sleeping/working room invite the out-of-doors in. The walls are covered with sheets, backtacked on. *Courtesy Martex*

UPHOLSTERER'S TAPE →

STAPLE →

This page is the wrong side of the second fabric panel. After the first panel was stapled to the wall, this panel was lined up directly over it (right side facing right side). This panel was then held temporarily in place with a few staples along its right edge. Directly over those staples was placed a strip of upholsterer's tape. Next, a row of staples was shot through the tape, the second panel, and the first panel. Turn back one page and see how the finished edge looks.

FOLD ALONG THIS LINE

Right- and Left-Handedness

While the staple gun is of neuter handedness, the directions given for its use in this book are given for right-handers. It doesn't always make a difference, but there are times when it does. For example, the directions for stapling fabric on walls are written working from the left side of a wall to the right. If a lefty tried to follow these directions he might find standing on his head the only way to do it. Instead, substitute right for left and vice versa.

Ironing and Pressing

Is it necessary to press fabric before stapling it? Pressing should be done, for example, when working on walls with gathered panels left loose at the floor. If a fabric is to be stapled on two sides only—as for a shutter panel or screen—all wrinkles that are at right angles to the direction of the stapling should be pressed out first. But when fabric is stapled with tension on all sides it should not be necessary to iron or press it beforehand. Knowing this can eliminate a lot of unnecessary fussing. If the fabric has been stapled on all sides and wrinkles remain, it:

1.   Probably has not been stretched tightly enough. Remove staples and tighten it. If *still* wrinkled, press while in place. If STILL wrinkled, it

2.   Most likely has a permanent press finish and a wrinkle has accidentally been permanently set.

Creases that result from being folded on the bolt—especially in some synthetics—can be stubbornly persistent. If this is the case, wetting, pressing, and restretching will lighten the crease, but nothing will completely eliminate it.

One Egypt-inspired print gives this room a unified, intimate air. The fabric was put on the wall using the backtacking method. John Elmo, Interior Designer. *Courtesy Belgian Linen Association*

Other folks might find more appeal in this dining room by interior designer Shirley Regendahl. Pure Belgian linen, woven into a bright cheerful plaid, is stapled to the walls and a bay window seat.

Finishing Touches

When gripper snaps became an acceptable fastener on the front of clothing, manufacturers began to make them decorative. Upholstery nails, too, come in a variety of ornamental styles, but no one has yet invented a decorative staple. Meanwhile there are many techniques, like backtacking, that camouflage or hide the staples. The strips of staples can be painted or stained with colored markers before they are loaded into the gun. Depending on the surface they are on, this may be enough to make them invisible. There are a great many trims that can be applied over the staples—wooden moldings, welting made from matching fabric or store bought, a galaxy of ready-made decorative braids and borders. All of these possibilities are described in more detail in other parts of the book. One of the ways that staples are hidden is by the wrapping technique—stretcher art and headboards involve staples only on the back side of the finished object—and even here there are ways to cover the staples. Overconcern for this is unnecessary. Even peacocks don't look so great from the rear.

Figuring Allowances

When figuring allowances for margins and hems it will be wise to remember what Goldilocks discovered when she tried beds on for size—just the right size is the best size. Too little allowance is difficult or impossible to work with. There should be enough extra so that the hand not using the stapler can get a good grip on the material to stretch it when necessary. But too much allowance is a nuisance. It gets in the way and may prevent a proper, tight fit. When possible, the directions for the project give the exact measurements that were used in making it.

Many prints are used in this pattern-on-pattern decor. Narrow strips of wooden molding, painted to match the baseboard, cover the staples at the ceiling. *Courtesy Belgian Linen Association*

An older bathroom is sparked up with flowered cotton curtains and walls. The staples just above the tile are covered with a row of double welting. *Courtesy Walls Uph*

3 Pads, Panels, and Screens

Large department stores' display departments spend a lot on materials generically known as composition board. These are panels of layered or compressed cardboard. Panels of foam sandwiched between covers of paper are often used too. Stapled onto these are covers of every imaginable material—felt, linen, straw, paper. These covered panels are used to line the walls of departments to give a boutique look, to form backgrounds for merchandise displays in cases and counters, or to create a colorful lining for an open armoire. All of these uses are applicable to the home. Because covering a panel with fabric is the easiest thing to do, this section is placed first.

Using Stretcher Bars

The method for covering stretcher bars is similar. Stretcher bars were originally sold only in art supply stores for canvas stretching. Now they are widely available—in art needlework departments to be used for mounting needlepoint and in yardgoods stores for specially designed graphics fabrics.

They come as small as 6 inches and go up to 60 inches in the lightweight and 72 inches in the heavy. They are sold in one-inch intervals. They are made of sanded, but otherwise unfinished, soft wood. Stretcher bars come in two thicknesses—standard and heavy-duty. The latter are found only in art supply stores and are intended for use where more rigid support is needed because of longer length or a heavier material to be supported. They are made of wider lumber and are more expensive.

Here is how to cover them simply and perfectly:

Assemble frame by pushing wood strips tightly together until corners form right angles. Check all corners against a T square or book to make sure they form a true square. Shoot a few staples across each of the joints on the back side.

Opposite page: Consider the possibilities in stapling ordinary kinds of coverings in unexpected places. Plain old homely kitchen toweling-by-the-yard was used to cover these cabinet doors. The bottom edge was stapled flat onto the under-edge of the door. The fabric was then folded back over itself, carried up over the front, and stapled along the top edge.

23

Stretcher bars have grooved, perfectly mitered ends that fit snugly together. They are usually sold in pairs and two sets are bought to form a rectangle or square. Elsewhere in the book other uses for stretcher bars are shown. They can be the base for diverse and wonderful home furnishing projects.

The focal point in this sunny alcove in a contemporary home is the puppy. Before his arrival it was the striking red, yellow, green, and blue graphic. Although it looks like an abstract painting on canvas, it is actually yardgoods on stretcher bars.

Right: To avoid a lumpy corner, staple one edge all the way to the end. Cut out a square of fabric as the photo shows. Turn in the raw edge and fold so that it falls exactly on corner edge. Staple down.

If fabric has a pattern that must be centered, lay the right side facing up and place frame over it. Mark where the corners should be with pins or lift the fabric and make chalkmarks on the wrong side. Cut out the fabric 2 inches larger than frame on all sides.

Lay fabric wrong side up on smooth surface. Wrap fabric over frame to the back and staple at center point of each side. Be sure that the fabric is very taut.

Starting from the center of the nearest side, staple to the corners, stretching fabric as you go. Reverse frame so that opposite side is near. Staple this side in the same way as the first. Do the same for the other two sides.

Finishing the corners. There is a tendency to stretch the material extra tight at corners. This can cause a distortion in the pattern. Check the right side frequently to make sure that it remains in line.

Turn the frame right side up and check to see if there are any places where the fabric is puckered or pulled too tightly. If so, remove one or more staples in the area, smooth fabric back to its proper position and then restaple.

Opposite page: Stretcher bar wall hangings are at home anywhere.
Courtesy Belgian Linen Association

24

Covering a Panel

In addition to stretcher bars, there is an unlimited choice of other panel bases. The ability to quickly and neatly cover a panel can lead down a path of endless home design possibilities. The variety of bases is almost as great as the selection of coverings for them. Using a base other than stretcher bars opens the choice to other shapes—circles, triangles, free form.

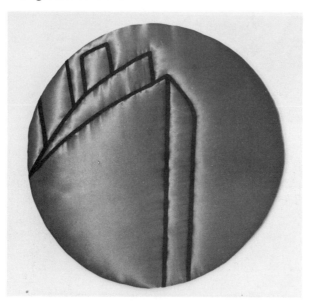

This art deco inspired wall hanging has an 18-inch circle of foam board as a base. The circle was cut out with a utility knife, padded with a layer of batting, and then covered with a circle of gray satin fabric. The "quilted" ocean liner design was transferred onto the surface of the covered circle with chalk. A desk stapler, opened up, was used to outline the design in staples. Black tape, glued over the staples, completes the hanging. The same staple-quilting technique can be used to outline a printed fabric design.

Supergraphic Panels

The manner of using the supergraphic panels in the photo opposite was inspired by the striking magenta and orange arc print of Scandinavian origin. The fabric was designed to be stapled to stretcher bars for instant dramatic art, but its graphic design suggested another way of using the fabric.

One section of the 60-inch-wide yardage was used to cover the center panel. Another section was split down the center at the point where the arcs meet and each section was stapled onto a panel. The two end panels were turned upside down in relation to the center panel, adding to the sweep of the pattern. The total effect is a vivid graphic design that puts zing into a plain white box of a room. Other prints form the stretcher bar textile collections that belong to the Knockout-Drop-Dead school of

This project was inspired by the striking magenta and orange arc print of Scandinavian origin. The fabric was designed to be stapled to stretcher bars for instant dramatic art, but its graphic design suggested another way of using the fabric. *Courtesy N. Erlanger, Blumgart Company*

interior design and can be used in ways beyond mere framing. In the chapter on walls, stretcher bar fabric is applied directly to the wall to form a mural and window shade combination.

Supplies and Tools 4 yards Materialize®, staple gun, staples in a length suitable to the thickness of the base, scissors. Center base is corrugated cardboard cut from a moped carton, outside panels are inexpensive wall paneling. Outside bases are 27 inches × 70 inches, center base is 54 inches × 70 inches.

Stretcher bars in this long length are not available except by custom order. Also, a flat base was chosen since the boldness of the color and design tend to make the panels visually jump out from the wall. Composition board or plywood are alternate bases for this project. Cut panels to size based on fabric width and design.

Base is placed on uncut yardage to make placement of size and design. Mark this with chalk.

After marking remove and split fabric in two. Cut fabric to fit base, adding a one-inch allowance on each side. This may seem too small a margin for error, but the stretching of the fabric as it is stapled increases the allowance. Accuracy is better than overabundance. (If a thicker base is used, allow for this when measuring.)

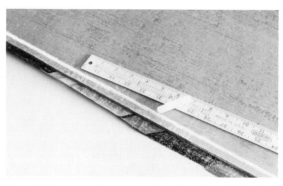

Place fabric wrong side up on work surface. Smooth it out and center base over it. It is important that the first side be stapled correctly. One way to do this is to mark a straight line with chalk along the base. Use the chalk line as a guideline for the fabric edge.

If one of the two long sides is a selvage edge, begin with that side. The selvage edge, combined with the guideline, insures that the fabric will be perfectly straight. Start at center and staple to each corner, stretching fabric while stapling. Reverse so that the opposite side is near. Smooth fabric under panel. At center place left hand on base with thumb on edge. With the other hand pull fabric tightly while the left hand pushes panel away. Without letting go, press panel down hard to anchor material while wrapping allowance onto top. Staple in place. Repeat at each end. Fill in with staples between center and ends, placing them close together. Check right side. If any adjustments are needed, remove the problem staples and restaple correctly.

Make a simple fold at the corner and staple it down. Check the right side to see that there is no distortion. Do the second panel the same way. This corner fold is done differently from stretcher bars because of the thinness of the base.

Covering an Old Desk

Part of the success of a project of this sort depends on locating a fabric with a design especially compatible with the area to be covered. These designs suit the drawer fronts so well that they appear to be hand painted. They came from three coordinating yardgoods—stripes, solid, and a paisley border print—all in the same color combinations. A cardboard was cut to fit each size drawer front to be taken to the store in the search for the most fitting fabric. It is impossible to depend on memory for matching. Furthermore, in the store, the choice is so great as to be bewildering. Time spent finding the perfect pattern will be worthwhile. Fabric can also be stapled directly onto drawers, but in this case the uneven finish would have shown through. This method also conceals the staples.

Close to fifty years old, this desk was a perfect candidate for rejuvenation in this manner. Although well constructed, its veneer finish was so badly damaged refinishing in the regular way was hopeless. Stained, chipped, and cracked, it was, to put it kindly, a candidate for the dump.

The transformation of this desk depends entirely on the simple technique explained in the last project—stapling a covering onto a pad.

Illustration shows areas that get new fronts cut from composition board.

Supplies and Tools Desk, fabric, paint, composition board, cardboard, utility knife, straightedge, white glue, small nails, staple gun and staples.

Finishing touches—The front drawer panels are attached with small brass nails, the back is screwed back on, the top panel glued on and a glass top for writing added. Bands of color picked up from the fabric are painted around the carved legs and the drawers are given new wooden knobs painted blue.

Chest is cleaned and damaged "gingerbread" removed from the front. A back panel of thin wood screwed in place is removed. After selecting the fabric, glossy enamel in a matching red is painted on those parts of the desk to be left uncovered.

Using the drawer fronts as pattern, trace their outline on a sheet of composition board.

Using a metal straightedge, run the knife blade lightly along the outline on the board. Do this several times to form a groove. Remove straightedge and use heavy pressure on the blade so that it cuts through all layers. If the scoring mark is deep enough, the blade will not jump the track but will cut cleanly. A muscular arm can cut through the board in one fell swoop.

Cut out according to design.

Place panel on wrong side of fabric. Staple in the center of one side, then at each corner, making sure the design remains in place. Fill in with staples between center and ends. Turn the panel so that the opposite side is near and staple this side the same way. Cover the front, side, and top panels this way.

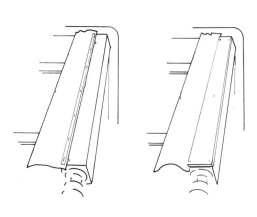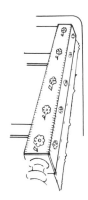

The front, inside, and back leg panels are done using a variation of the backtacking method. Staple the fabric piece for the front leg directly to the desk, wrong side up. Staple a strip of upholstery tape over it. Attach composition board panel to desk with small nails or long heavy-duty staples. Turn fabric right side out and bring it over the front and around to the side. Staple the fabric flat on the side of the desk. (The raw edges and staples from front and back legs will be covered by the wrapped side panel.) Cover the inside front leg the same way, wrapping the fabric inside the kneehole.

Wrap the back leg, clipping with a scissors where necessary. Turn under a hem along the back edge of the leg, glue, and pin it until dry. The steps for a project like this are done as an upholsterer covers a chair, with each succeeding piece of fabric layered over the last.

Other Ways to Use Panels

The same covered-panel technique can be used in many ways in interior decoration whether it is for a city apartment or a country home; whether the decor is modern or period. Slabs of covered foam building insulation are a quick way to create dimensional graphic wall art. Although the foam is extremely air filled, staples grip in it well enough to hold all materials except for heavy weights such as canvas. It works very well with knit fabrics.

A more traditional treatment for a stapled panel is using it to enhance a door. Take cardboard panels cut to size along to the store and try fabrics on for size.

An architecturally blah room can be given the interest of a paneled room by covering panels in a scale and design worked out to a custom fit. The panels can be trimmed with wood moldings.

Lewis Forman Day published his decoration combining a Greek key and floral motif for a well-dressed Victorian door in 1887. It might be adapted for a door today using pads wrapped in printed fabric. A handsome paneled door is not necessary—a stock flush door can have wooden molding applied around the panels. From "The Planning of Ornament." *Courtesy New York Public Library*

C.F.KELL,LITHO, 8, FURNIVAL ST HOLBORN, E.C.

Mounting Other Things on Panels

Using either stretcher bars or a solid panel base, a beach towel can make a colorful wall hanging. The bright colorful designs now available in towels give a poster art look especially suitable for a child's room or a game room.

Quilts or fragments of quilts, Oriental rugs, crocheted antimacassars or lace tablecloths—all are good candidates for turning into decorative wall panels.

Making a Picture Frame

The technique just shown for covering a pad is the same one used to make a handsome picture frame. Considered this way—two covered pads, one with an opening in the center, sandwiched together—it sounds simple. It *is* simple to do and the result will be an attractive frame custom designed both in size and material for a favorite photo.

The interesting collection of frames for this rogues' gallery was made from bits and scraps found around the house.

Work out the dimensions for the frame based on the size of the photo to be framed. Although the dimensions of the photo may be square or rectangular, the subject matter should determine the size and shape of the opening. Perhaps a round or oval opening will form the perfect frame for the subject. The opening in this frame was centered and the dimensions chosen for a 5×7-inch photograph.

Supplies and Tools Cardboard cut from a corrugated carton, white glue, a remnant of fabric, and masking tape. If glass is to be put into the opening a piece the proper size will be needed, plus household cement to glue it on the cardboard. If an easel back is to be attached to the back of the frame, two round-head brass fasteners are needed. Tools needed are scissors or utility knife for cutting cardboard and a desk-type stapler with an anvil that can be swung out of the way. A staple gun may be used—it is easier to work with—but the staple length must be short enough not to perforate through the cardboard and front covering.

The very same technique can be used to make large frames, substituting plywood for the corrugated cardboard. Or use existing frames—either inexpensive five-and-dime store ones or junk shop derelicts.

With a utility knife, cut two rectangles from corrugated cardboard, each 9½ × 12 inches. In the center of one cut an opening 6 × 4 inches. Use a straight edge to measure and cut very accurately. Wrap edges in masking tape.

Cover back half first to gain practice. Line up fabric following the design if there is one, or the grainline if not. Grainline is especialy important when using a distinctive weave such as burlap or raw silk. Start stapling at the center of one long side, stretching fabric toward corners as you go. Back to the middle, repeat, stapling and stretching the fabric toward the other corner. Fill in with closely spaced staples. Check front to see that it is in proper position. Staple opposite side, then two ends.

There should be four dog-ears at the corners. If the excess fabric does not stand up freely, adjust staples so that it can. The bulk on the corners should be reduced as much as possible so that when completed, the two halves of the frame can press snugly together. With a sharp scissors laid flat against the cardboard, snip into the corner as shown.

Now hold scissors vertically over the point just snipped and cut down to meet the last cut. Remove this triangle of fabric, grasp the tip of the dog-ear, and pull it down toward the center of the rectangle.

Staple it firmly down. Repeat this at the other three corners.

Cover the front half of the frame the same way.

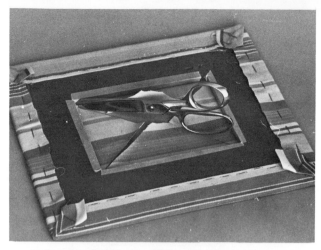

With sharp point of scissors, slash fabric from center exactly into each of the four corners. Staple these triangles to inside, pulling fabric as snugly as possible without distorting the design. Check right side often to make sure pattern is still straight.

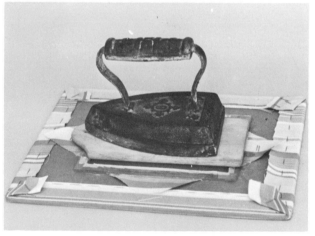

Spread white glue generously on both sides of the four inside corners. Use pins to hold fabric in place until dry. This step is tricky as there is a bare minimum of fabric to cover the cardboard. Work carefully, but do not fuss or the fabric will fray.

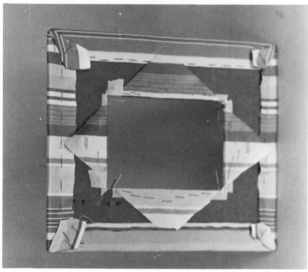

Use household cement to glue glass in place and weight it down until dry.

The frame can be made so that photos may be changed. To do this, glue a strip of cardboard over the bottom edge of the glass to prevent photos from slipping down.

Cut a four-sided triangular shape from the corrugated cardboard to make the easel-back. Follow the look of the one in the photo. Exact size is not important. Cut it so that the corrugations run vertically. With a ruler, make a depression along the length of the long side about an inch from the edge. Line up the depression facing down along a sharp counter or table edge and bend the cardboard. Flop it over and flex it in the opposite direction. Fold it back and forth until a deep crease is made. Staple the fabric to the long side. Spread both sides of the cardboard with white glue and wrap fabric snugly around, folding the top edge under into a narrow hem. Press fabric into the crease previously formed and prop it up until dry. When dry, trim bottom edge so that the frame will stand at a slight back-tilted angle.

Position the easel-back on the frame back. With a nail or sharp scissors, puncture two holes through both layers and insert two round-headed brass fasteners. Weight down until glue dries. Insert photo if one has not already been taped into place. Run a generous strip of white glue near the edge of both halves. If photo is to be removable, slip a piece of waxed paper between the two layers at the top of the frame and glue only the other three sides. When dry, remove wax paper.

Screens

Until recently folding screens were seen mainly in the movies. Heroines in Westerns and backstage dramas used them for demure costume changes. Historically screens were decorative items, perhaps covered in rich brocade, their function limited to serving as a background for other objets d'art. Currently their popularity in the interior design world is due not only to their decorativeness but to the great variety of functions they fill. They can screen a dreary view, serve as a draft stopper, provide privacy, block a cluttered work area from sight, hide an unused doorway, or divide a room. Radiator covers waste fuel; a screen serves to disguise the radiator but does not block the flow of heat.

Since screens have few structural requirements for strength and since they get little wear and tear, materials for them are not limited to wood. Composition board, insulation panels, even a corrugated bicycle carton can supply the material for a four-panel standing screen.

Screens don't have to be made from scratch. Inexpensive lightweight ones are available in a variety of sizes and number of panels from the large mail order catalogues. Old ones can be found in secondhand furniture stores. Recover them exactly as they were or adapt some of the methods in the following pages. A solid fabric may be replaced with sheer. A flat screen may be padded or covered in quilting. Wood frames may be painted and left uncovered with new center panels of pleated fabric stapled in place. The following projects give instructions for different ways of finishing the edges.

Restoring a Victorian Wood Screen

Supplies and Tools Wood frame screen, fabric upholsterer's tape, scissors, iron, white glue, staple gun and staples.

This photograph shows the general condition of an oak screen immediately after it was rejected by an antiques shop. 179 rusty tacks and a few crumbling remnants were all that remained of two earlier fabric covers. The frame was cleaned, retouched with stain, and waxed. The project is included to show a staple gun transformation that can be used for many other projects.

Opposite page: The frame has no right or wrong side. For consistency's sake, one side was selected to be the front. A striped fabric that looked good on both sides was chosen. Any woven design would be a good choice. While the majority of printed fabrics have a reverse side that literally looks like the "wrong side," there are exceptions—this printed striped fabric is one of them. *Courtesy The Stock Market, Westport, Connecticut*

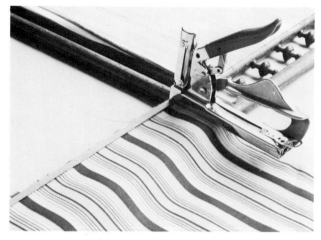

Measure, cut, and press fabric. A selvage edge serves as hem on one side. The other side is folded and pressed into a narrow turned-under-twice hem. The tautness of the panel keeps this fold closed without need for sewing.

Lay screen flat on work surface, back side up. Place first fabric panel on screen wrong side up. Flop panel completely over, keeping top edge against top frame. (This edge will be backtacked.) Position fabric and secure it with a few staples. Line edge with a strip of upholsterer's tape and staple across from end to end. Flop fabric over. Staples are hidden.

Stretch fabric tightly down to bottom of frame. Do not stretch fabric widthwise or stripes will be distorted. Staple bottom edge in place using flat stapling.

Cut a strip of upholsterer's tape the width of the frame. Trim away extra fabric so what remains is no more than twice the width of the upholsterer's tape. Run white glue along both sides of strip. Wrap the fabric around tape and press down. Keep in place until glue dries with baste staples or masking tape. Cover the other panels the same way.

Chiffon-covered Screen

This screen was built to fit a piece of mauve chiffon found at the remnant counter of a dressgoods shop. Because of the bold art deco design the size of the screen was planned for both dimension and placement, and to get maximum use of the available fabric. The fabric was folded in half and then adjusted so that the pattern of the two halves would line up perfectly (pattern repeat). Fortunately only a few inches of fabric had to be wasted. The two halves of fabric were then laid side by side and considered in regard to the best width for the panels. The best proportion for each panel was established to be 16×72 inches.

Light filtering through this screen brings to mind stained glass from the art nouveau period. It can be copied very easily.

The frame is painted the same color as the background of the fabric using acrylic paint from art supply store.

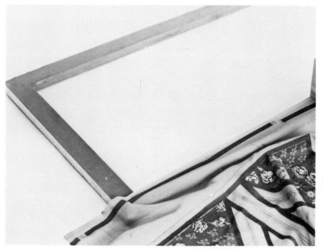

Because placement of fabric is important, and because panels are already cut to fit the frame exactly, the usual step of centering frame over fabric is eliminated. Instead of beginning at center, begin at top as placement of design on fabric must be lined up here. (Amount of stretch from stapling would be unpredictable.) Staple all the way to bottom, stretching fabric as you work. Turn panel and staple other side starting again at the top. Staple ends.

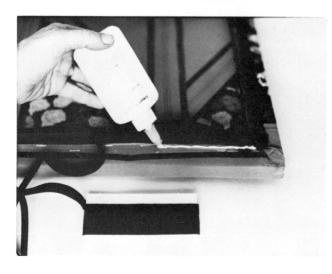

Fold corners into dog-ears and finish corners the same way as for the picture frame. Trim off excess fabric from all four sides. Cover the remaining panels the same way. Run a bead of white glue over raw edge and cover with a strip of seam binding. Allow to dry.

Line up panels on work surface and attach hinges. If half-fold hinges are used, mount them on the back surface for a backward fold and on the side for a forward fold. Or use universal hinges that allow panels to swing both ways. Mount them on the inside edge.

To give additional height to the finished screen decorative wooden knobs available at lumberyards or hardware stores can be mounted on the bottom. These or furniture glides will give the fabric wrapped over the bottom edge protection from wear.

To emphasize the light-filtering quality of the sheer fabric, an open-frame construction with a single layer of covering was used for the last project. There are several other methods for constructing folding screens using a solid base.

Supplies and Tools 2½ yards of 45-inch-wide fabric, white glue, 15 yards of hem binding or single-fold bias tape, paint and brush, staple gun, and staples. Stretcher bars make a perfect frame for this kind of screen. Or with more effort but at a lower cost, a frame can be constructed. This frame was made from 1×2-inch lumber. The corners were cut in a simple miter box using a hand saw, but they can also be made with a butt joint. The corners were fastened with a combination of white glue and 3-inch nails hammered in at angles from either side of the corner. After drying, it became noticeable that the frames bowed out in the center (a possibility with a frame as long as this), so center braces were glued and nailed in place forcing the frame to straighten. Because of the sheer and delicate nature of the fabric the frame was sanded and the corners slightly rounded off by sanding to prevent a sharp corner poking through.

Folding Screen with Fabric Hinges

Henry B. Urban, urbane upholsterer to the carriage trade in Manhattan, taught me to make this elegant upholstered screen with hinges as intriguing as an ancient Chinese puzzle. See page 44. No magic is required to duplicate it, just his directions. Although Mr. Urban's shop uses tacks and hammer in the fine tradition of the English school of upholstery, the same techniques are adapted here to the staple gun.

Whereas hardware hinges would add a jarring flash of metal, the self-fabric hinge with its interwoven pattern is an integral part of the screen. Like a universal hinge, it flexes in both directions. But good looks and performance aside, it eliminates the cost of the hardware and can be made from scraps of fabric. Note: As the hinges are attached first thing, they cannot be added as an afterthought.

Supplies and Tools 2 insulation panels 24 inches by 8 feet, 8 pieces of 16-foot-long ⅞-inch lattice, white glue, scissors, saw, staple gun and staples. To cover a pair of double panels 82 inches high by 12 inches wide requires 6 yards of 55-inch-width fabric. If fabric is not wide enough to provide a 3-inch-wide strip to line hinge edges, an extra length of fabric will be needed. (When totaled, the supplies for this project are not inexpensive. But considering the yardage that would be required for triple-pleated draperies for the same area, and the fuel economy and comfort provided because of the insulation, the cost becomes realistic.)

Folding Screen with Fabric Hinges (cont'd)

These tall screens covered in terra cotta linen are a handsome and practical way to treat a large glass expanse. Choice of a solid color accents the criss-cross pattern of the self-fabric hinges.

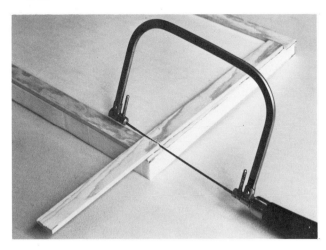

Insulation panels are cut to size using utility knife or saw. Cut lattice strips to form an outline around foam.

Glue lattice onto foam with white glue. Weight down with books while drying. If lattice does not true up to foam edge, force it to by shooting staples in. Remove staples when dry. Attach lattice the same way to both sides of the panels.

Cut narrow strips of fabric the length of panels. Wrap and staple it smoothly around both sides of the edge to be hinged. Do the same to a second panel, then line them up side by side on the work surface.

From the fabric cut thirteen rectangles 5 × 13 inches. These are the cloth hinges. (Any size and number of hinges may be used.) Fold each one in half crosswise as in the photo and press a ¼-inch hem across the short ends. (A lining may be added; it adds strength and a finished appearance.) After pressing hinges, place them on alternate sides of the panels. Staple as shown.

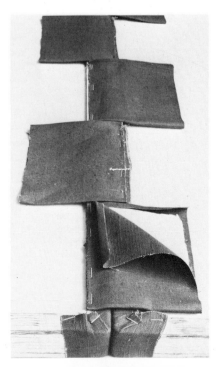

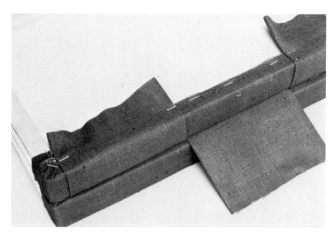

After stapling hinges, separate panels. Fold the right panel on top of the left. Wrap the hinges from the bottom panel and staple them onto the top one. Turn both panels over and staple hinges from the bottom panel onto the top one. Be sure that the edges of the two panels stay lined up.

Trim extra fabric off hinges. If heavy fabric is used, hammer a few upholstery tacks through hinges into lattice for extra strength.

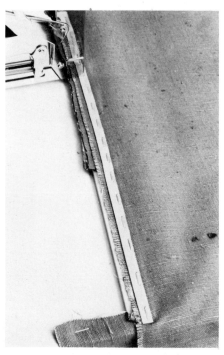

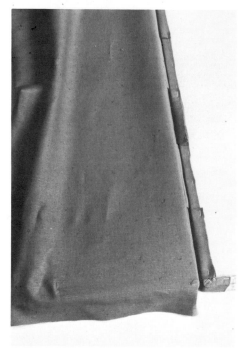

Cut a piece of fabric 4 inches longer than the panel, wide enough to wrap around both sides. Use upholsterer's tape and backtack it on the length of the panel at the hinged edge.

Fold ends as though wrapping a package. Glue and baste staple the ends. Turn under the raw edge and flat staple along the hinged edge. Glue braid or self-welting over the staples. Or, as Mr. Urban would do, hand stitch the seam using a curved upholstery needle.

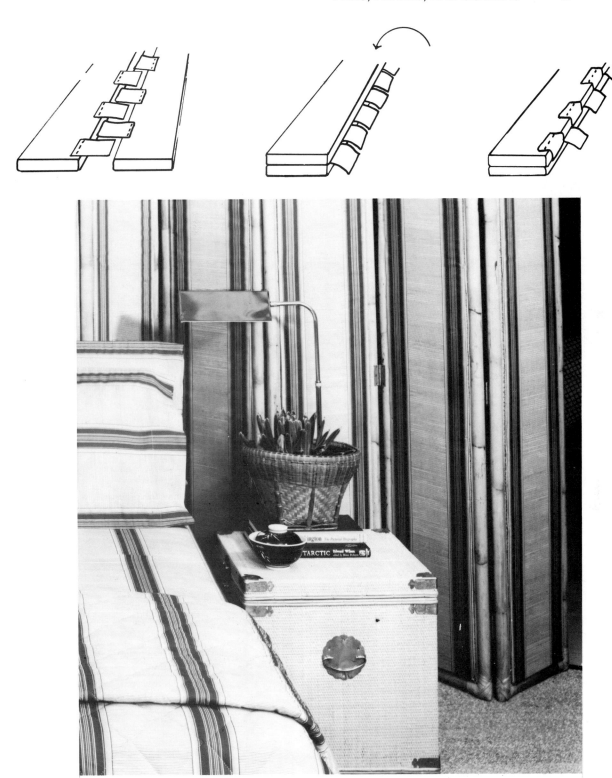

Natural bamboo outlines a narrow, multipaneled screen covered with striped sheeting. Bedspread, sheets, and screen make a tailored, attractive corner. *Courtesy Utica for J. P. Stevens*

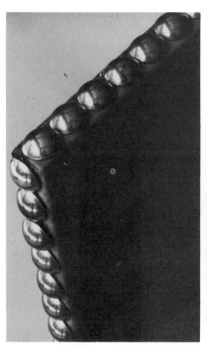

With this method, the covering is brought separately from front and back and stapled to the edge.

The edge may be covered in a variety of ways. Jumbo brass upholstery tacks were used on this screen and completely cover the staples and raw edges. Colored thumbtacks may be used for a more casual look. Or braid may be glued on.

Self-welting makes a less noticeable finish.

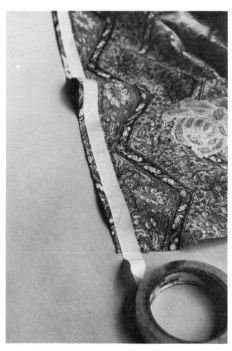

A quick way to make self-welting is to stick masking tape to the wrong side of fabric. Allow a quarter-inch on each side of the tape.

Fold over and crease along tape edges, then glue.

Another Way to Cover a Screen

The mauve chiffon screen demonstrates how to cover a screen with a single layer. The Victorian screen demonstrates how to use a single panel in the center of a screen. The last project describes the method for using one piece of fabric wrapped around both sides of the panel. There is another simple way to cover both sides of a screen. This method uses separate pieces for the front and back. Perhaps there is a limited amount of the covering available—a plain muslin lining will do for the back. Perhaps the front covering is expensive—the cost can be kept down by using a compatible but cheaper covering for the back. If the screen is to be used at a window a neutral color facing the street may be required. A reversible screen with a light summer side and a warm winter one makes good sense, too.

Stretcher Bar Screens

Because they are neatly mitered and sanded, stretcher bars lend themselves to quick and easy screen making. The frames may be left plain or given a coat of wax, stained or painted; they need not be covered. Refer to the directions given for re-covering a Victorian wood screen and apply to stretcher bars for a contemporary look. A superfast way to make a tailored-looking screen is to staple on upholstery webbing. For additional interest, the webbing may be woven. Screens may be hinged quickly by drilling holes through the frame of the adjacent panels and lacing them together with leather thongs.

4 Walls

━━ ══ ━━ ══ ━━ ══ ━━ ══ ━━ ══ ━━ ══ ━━

Using Fabric on Walls

Walls covered in fabric always look expensive and elegant. This is true whether the fabric is precious cut velvet or irregulars of inexpensive muslin sheets. Centuries ago walls were hung with tapestries and brocades and rugs, not only for the richness of texture and color but also to provide comfort in drafty castles. Even the simplest peasant cottage used textiles as an insulation material as well as for decoration. Today there is a wealth of color, pattern, and texture available such as was never dreamed of in times past, but the reasons for choosing fabric as a wall covering can still be practical ones.

While paint is a less expensive covering than fabric, cracked and peeling walls—or walls covered with damaged paneling or unwanted tile—can be covered with fabric more easily than they can be restored to paint-taking smoothness. If such is the case the saving in labor can be significant.

What about wallpaper? There are several reasons to choose fabric-covered walls rather than wallpapered ones. Wallpaper—especially in quality designs and materials—can be more expensive than fabric. Preparation of walls for papering is time consuming, as a smooth surface is a must. Stapling fabric to a wall is far easier than coping with messy paste, air bubbles, and crookedly applied paper panels. Fabric also has the advantage of being removable, and can be taken along to another apartment or used again for another purpose.

Holes left by the staples when fabric is later removed are no problem. There is no need for plastering or spackling. Give the wall a coat of paint—regular latex wall paint fills the staple holes and leaves a smooth surface.

Because of the sizing present in new fabrics and the soil-resistant finishes that are

What might be one person's idea of heaven, this is actually the artist Colette's Pearl Street loft in New York's SoHo neighborhood. She creates these silken environments using a staple gun. Notice the strings still attached to the parachutes trailing to the floor. Courtesy Eugenia Cucalon Gallery. *Photo by Al Mozell*

widely used, maintenance does not have to be a problem. Vacuuming keeps them fresh for years. A washable fabric can be spot washed with detergent; other fabrics can be touched up with cleaning fluid. Of course areas where soil is likely to be present—such as kitchens or nearby walls—don't take well to this application. And if there are little hands around the house prone to peanut butter smearing, better stick to wallpaper for the time being.

While padding under fabric is most often added for the beautiful soft appearance it gives, it has other reasons to recommend it. It helps to deaden sound from inside or outside the building and from the apartment next door. It can be a temperature insulator as well as a sound insulator. Applying a layer of padding between wall and fabric is not difficult to do. The techniques shown in this chapter are not limited to fabric on walls. Sisal or straw mats, leather (real or fake), shiny vinyl, gray flannel—all are possibilities.

Using Sheets to Cover Walls

Design excellence runs rampant in the sheet industry. Top designers from the fashion world as well as the home furnishing field have applied their talents to the sheet market. Even a handful of show business personalities has gotten into the act and added a colorful touch. Fine artists are represented, too. The heirs of Picasso have made available adaptations of his drawings; museums allow copies to be made from their textile collections. Although "white sales" are still held in bedding and linen stores, they feature a riot of color and pattern and very little that is white. A Macy's sale catalogue recently pictured thirty-five different sheet patterns, many of them available in several different color ranges. Today's linen closet can harbor bold geometrics, dark solid colors, cartoon character prints, conservative checks and stripes, and jungle animals in every conceivable species. As a result of this, sheets have come out of the closet and made themselves at home on the wall of every room in the house.

All this proves that good design doesn't have to be expensive. Inch per inch, sheets are inexpensive fabric. In addition to the sales already mentioned, many sheet patterns are eventually available as seconds or irregulars—and then marked at about half the regular price. The imperfections rarely affect their decorative use.

Another great thing about decorating with sheets is that they often come with matching blankets, comforters, spreads, draperies, and ruffled pillows. Often, too, they are done in not only matching but coordinating designs—stripes planned to go with checks, geometrics with florals. This takes the uncertainty out of mixing and matching.

One of the best features of sheets is their big size. One flat king-size sheet will cover a wall 9 feet by 9 feet. Three twin-size sheets will cover a wall 16 feet long. Another plus is that since the flat sheets are finished on all four sides (woven selvages on two sides and hems top and bottom) they can be used for many purposes as is.

Zebra and tiger stripes cover the bed and wall in this city bedroom. *Courtesy Burlington Mills*

Opposite page: A single sheet is stapled to the wall behind this charming old iron bedstead. Ready-made sheer draperies and a quilted spread create a charming nostalgic scene. Not surprisingly, when the bed is turned back, sheets and pillow cases feature the same flower print. *Courtesy Dan River*

Racing stripes wrap around the walls of this bathroom. Fabric can be used in a shower area if thoroughly waterproofed with a sealer. Clip-on shower curtain hooks turn one flat sheet into an instant shower curtain. *Courtesy Springmaid Sheets*

See How Easy It Is. Flat Stapling One Sheet to a Wall

The staple job on this tall wall (opposite page) was done by one person working alone and so required many trips up and down a ladder. Try to have another pair of hands to help out when working with a panel as large as this.

Open the hems at both ends of the larger sheet and press flat. Position the sheet on the wall with baste staples. Step back to see if design is level and placement is pleasing. If not, remove baste staples and reposition. Start stapling fabric to wall at center of left side. Stretch the fabric up while stapling toward ceiling and down while stapling toward floor. Smooth the sheet toward the right edge, making sure that there are no wrinkles and that it is as taut as possible. Staple the right edge the same way the left was done—start from the center, stretching upward and down while stapling.

Staple the top as close to the ceiling as possible. Staple the bottom close to baseboard or floor.

Use a single-edge razor blade to trim excess fabric from top and bottom. (Used this way, blades get dull very fast. Large quantities can be bought cheaply at paint stores.)

Measure and cut the border trim for the left side of the wall from a twin-size sheet. Turn under raw edges into a hem and press with an iron. Baste staple the border loosely in place. Staple top edge to wall. Run a bead of white glue along the turned-under hem on each side, pat it into place, and keep it there until dry with baste staples or a row of masking tape. Because of the busy design of the sheet used, the staples are almost invisible. Numerous ways to hide them are described throughout the book.

Supplies and Tools One king-size flat sheet, one coordinating twin-size sheet, white glue, staple gun and staples, single-edge razor blades. See sheet sizes chart. A smaller-size sheet may cover the entire wall of a smaller room.

Opposite page: Beautiful big roses make a dramatic spot in this all-white room of lofty proportions. With help from a striped border, one king-size sheet covered the wall completely.

Another way to use one sheet is shown in this photo. A queen-size sheet in a dramatic gardenia print designed by Luis Estevez is centered behind the headboard of a single bed. The portion behind the bed is stretched and stapled directly to the wall. The rest of the sheet is draped into loose pleats on either side of the bed. The staples attaching the sheet to the wall are hidden by the draped fold. *Courtesy Dan River*

The decorator who planned this wall treatment got a lot of mileage from the western landscape printed on the sheet. One twin-size is wide enough to cover the wall behind a double bed and long enough to reach floor to ceiling. In this setting, an existing wooden frame surrounds the sheet like a painting. The same effect can be copied with wooden molding. Follow the directions just given for stapling up the king-size sheet—position the sheet with baste staples, then stretch as you staple sides, top, and then bottom. *Courtesy Burlington Mills*

Using Fabric as a Mural

Before a design is selected, it is important to work out measurements as well as pattern. Some graphics are printed vertically, some horizontally. Using a 47-inch-wide cotton by Materialize, it was possible to fabricate a window shade without making a break in the pattern. (See chapter on windows for directions for making the laminated shade.)

Supplies and Tools 4 yards were used for a 12-foot-long wall. (This design is printed horizontally so yardage is determined by the width of the wall. The shade is made from section cut out for window opening so no extra yardage is required for it.) White glue, 3½ yards of double-fold bias tape, 3 yards single-fold bias tape, scissors, paint for staples, staple gun and staples.

Supergraphic textiles, intended for stretcher bars, can, when stapled directly to the wall, become unique murals. The superscale strawberry print in brilliant red, green, and blue gives a dramatically bold effect to a plain window wall. See pages 156 to 157 for shade directions. *Courtesy of N. Erlanger, Blumgart Company*

So as not to distract from the design, moldings or decorative trim are not used. Instead the staple stacks are given a coat of paint before being loaded into the gun. They blend into the design. A staining-type marker can be used to go over visible staples if only some will show, or if a multicolor background is used.

Mark center of fabric and center of wall. Baste staple the fabric on the wall.

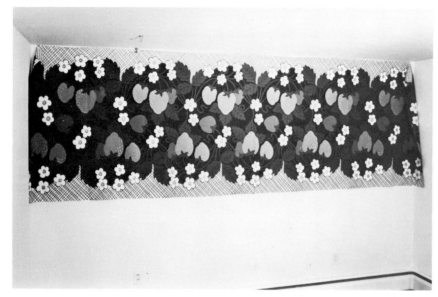

Stand back to check design placement.

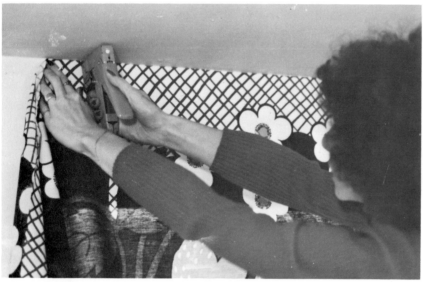

Begin stapling fabric to wall from center. (One selvage is along ceiling.) Work toward corner, smoothing and stretching fabric as you work. Ignore window opening—it will be cut out later. Remove and adjust staples as necessary in order to get the whole wall covered smoothly.

Use a yardstick to chalk lines outlining edge of window opening. Baste staple a few inches back from chalk line. If window has frame, it can be incorporated into the graphic design by painting it the same color as the fabric ground.

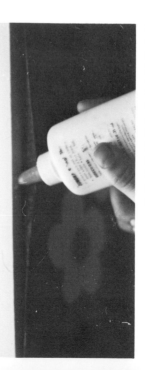

Staple fabric to wall around window opening. There may be a metal strip outlining window just beneath plaster that prevents staples from penetrating. Cut out fabric and glue around the opening. When glue has dried remove baste staples. Glue a strip of double-fold cotton bias tape all around window opening. If the window has a wooden frame, cover raw edge and staples with a strip of single-fold tape.

This is the flat stapling method. Pattern matching is easiest with it. Self-welting glued over staples makes seams almost unnoticeable.

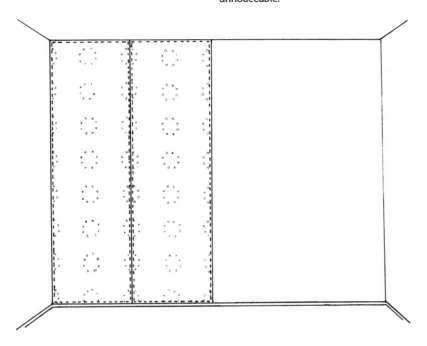

Covering All the Walls of a Room with Fabric

The method for covering all the walls of a room builds on the same simple steps used for the strawberry mural and the single sheet wall. Although all of the factors that contribute to a professional job are described in detail, don't confuse detail with difficulty. Stapling fabric to the walls of a room is an easy thing to do.

This small room in an old New York brownstone is liberally frosted with fabric. Designed for Quadrille by Jay Crawford and Anthony Tortora, this fabric is a perfect choice. With all surfaces of the room in harmony, the walls seem to be pushed back. *Photo by Richard Champion*

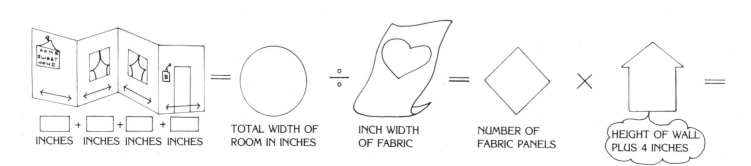

INCHES INCHES INCHES INCHES   TOTAL WIDTH OF   INCH WIDTH   NUMBER OF   HEIGHT OF WALL
                                    ROOM IN INCHES   OF FABRIC   FABRIC PANELS   PLUS 4 INCHES

Before Starting

Figuring Yardage When Using a Patterned Fabric

Always consider the fabric in terms of panels—even if the panel is a 9-foot-wide king-size sheet.

If the fabric has a pattern repeat, this factor must be taken into account when figuring yardage requirements. A pattern repeat is the number of inches involved before the identical design begins again. The repeat of a pattern refers to its lengthwise design. Most decorator fabrics have horizontally balanced patterns that match up perfectly along the selvage edges. This horizontal or widthwise repeat is known as the way the pattern "reads." Two examples of exceptions are border prints and asymmetrical plaids.

The size of a pattern repeat can range from a fraction of an inch up to the width of the metal roller that prints the design onto the fabric. A 36-inch repeat is unusual, but many typical decorator textiles have repeats that range between 22 and 28 inches.

Some repeats are not obvious when a small section of fabric is examined. The eye can be tricked by designs made up of small flowers—their individual smallness may be part of a large repeat and this will not be evident until a large sample is examined, preferably from a distance.

It is usually true that the larger the repeat, the more waste will occur—and so the greater the expense. But this is not always the case. The important factor is the combination of the length of the area to be covered with the number of times the pattern repeat will fit exactly into that area. For instance, a fabric may have a 36-inch repeat, but if the area to be covered INCLUDING TOP AND BOTTOM ALLOWANCES is 72 inches, there will not be one inch of waste. In this case, two repeats, or 2 yards of fabric, will fit exactly. If, however, the 36-inch repeat is for an area requiring 79 inches, three repeats will be required. Each panel will have a waste of 29 inches. If a large area is involved, a large amount of waste could result. (There are many good uses for this extra—it can be used for welting and trim, pillows, curtains, and so on.)

This relation between wall height and pattern repeat may determine which fabric is selected. One fabric, though priced more per yard than another, because of its appropriate spacing of repeat may work out to be less expensive than a lower-per-yard fabric with a more wasteful repeat. Solid colors, of course, have no waste and neither do stripes, regardless of whether they are bold awning size or narrow mattress ticking.

An interior designer who worked out pattern repeat yardage requirements based on square root and other mathematical computations got a fresh perspective from a little old lady who "takes in sewing." Like the grocery game played on the TV quiz show—how many boxes of food can be bought without going over a certain sum of money—her game asks: How many pattern repeats will fit into the height of the space without going under?

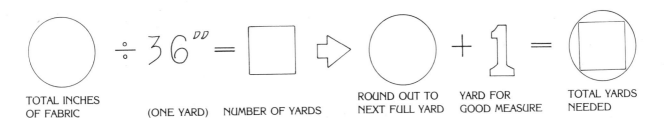

TOTAL INCHES
OF FABRIC   (ONE YARD)   NUMBER OF YARDS   ROUND OUT TO   YARD FOR   TOTAL YARDS
                                                     NEXT FULL YARD   GOOD MEASURE   NEEDED

Mandarin-red sheet with a decorative border is used as a hanging for this Chinese bed. It looks so perfect, perhaps the bed was carved to order. *Courtesy Wamsutta*

Figuring Yardage When Using Sheets

Sheet dimensions are given in what is known as cut size. This means that the width of the sheet is exactly as stated from selvage edge to selvage edge. The length given, however, is measured before hems are turned at both ends. The bottom hem is 1¼ inches, the top is 5 inches. (Occasionally the hem allowance is divided equally between top and bottom.) Often the top hem is sewed on as a decorative border. With printed sheets there is no guarantee that the repeat will fall in the same place on all sheets. Although they usually start and stop at approximately the same point in the pattern, there will be more variance in trying to match up different-size sheets. This should be no problem, as the length of a twin-size sheet is 8½ feet, and this should provide enough allowance for matching patterns. If the room is super high, king- and queen-size sheets give another 6 inches of length. Sheets are not printed to match selvage to selvage, but the waste usually amounts to no more than 6 inches.

Sheets can be used widthwise if the dimensions work out better that way, provided the pattern is not one that obviously runs lengthwise. (A roomful of roses lying on their sides would not be attractive.) Whichever way is used, place the sheets so that they all go in the same direction. Another design to watch for is the panel print. This is worked out so that a single motif is centered on the sheet. This can be most effective, but only if planned with careful spacing in mind.

Standard Sheet Sizes

Twin   66 × 104 inches = 5'6" × 8'8" = 6,864 square inches = 39 × 75 mattress
Full   81 × 104 inches = 6'9" × 8'8" = 8,424 square inches = 54 × 75 mattress
Queen 90 × 110 inches = 7'6" × 9'2" = 9,900 square inches = 60 × 80 mattress
King   108 × 110 inches = 9'0" × 9'2" = 11,880 square inches = 76 × 80 mattress

Twin is also known as single, full as double, king as dual (two extra-length twin beds together).
Standard-size pillowcase cut size is 42 × 36 inches.
King-size pillowcase cut size is 42 × 46 inches.
Drop (height from top of mattress to floor) is usually 20 to 21 inches.

Plumb Lines

The sure way to eliminate frustration from fabricking walls is to understand and use the plumb line. A plumb line is a true vertical line established by gravity. It must be honored, whether the house is a literal "lean-to" or a grand old Colonial. A house may have been built with all walls at right angles to each other but shifted over the years leeward to the wind; new luxury apartment buildings may have been carelessly built out of plumb with no prospect that they will straighten as they age.

How does this apply to someone with fabric and staple gun in hand eager to get started? If a fabric is lined up with a room corner that is not plumb, the pattern on the fabric will begin to creep up or down the wall depending on the direction of the slant. This will not be very noticeable with solid-color fabric, but with patterned fabric it can be

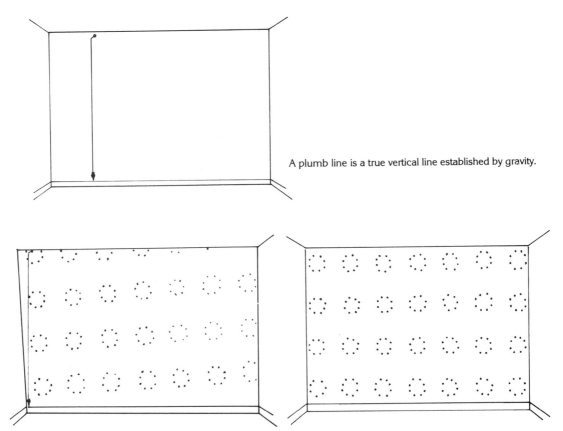

A plumb line is a true vertical line established by gravity.

This wall was covered without first checking the left corner for plumb. Fabric was lined up on left selvage edge and with each panel the creep upward is more noticeable. A second mistake was not to set the top motif of the pattern as though it were framed by the ceiling.

Plumb perfect. Also notice how the top row of the design is spaced at the ceiling.

distractingly obvious. The patterns that show this the most, and are therefore the hardest to work with, are those with a horizontal design. When working with designs that have a strong horizontal design (plaids, for example), the level of the ceiling is another equally important factor. The true level of the floor is not important, as the line is broken up by furniture or draperies. When working with fabrics such as checks, plaids, or horizontal stripes it is necessary to consider the levelness of the ceiling. There can be a double problem if the ceiling is not level and one or the other corner of the wall is out of plumb. In this case don't attempt to use one of the types of designs mentioned. If the plumb and level problem does not become evident until fabric is bought and ready to hang, a compromise between ceiling level and plumb line is the best choice. Geometric designs are best avoided when the straightness of walls and ceilings is in doubt. A random pattern with a small repeat is the best choice. When doors and windows are out of plumb, outlining them with a contrasting molding will accentuate the problem.

Making a Plumb Line

Tack a piece of string that has been rubbed with chalk to the top of the wall. Tie a plumb bob or a heavy object such as a hammer to the other end of the string so that it nearly reaches the floor. When the bob has stopped moving, anchor the string to the wall with one thumb. Use the other hand to snap the string. It will leave a "plumb" chalk line on the wall. (On white walls use colored chalk.)

Seaming

Another way of installing fabric is to seam the panels together on a sewing machine so that an entire wall can be installed in one piece. The advantages to this method are that fewer vertical seams are required, stitched seams may be less visible than when done with backtacking, and fewer staples will have to be used. On the other hand, for a person working alone one large piece is more difficult to handle, a sewing machine must be available, and the fabric, secured in fewer points, may sag or droop away from the wall.

Color of the Wall May Be a Factor

Many fabrics, especially those light in color or weight, let the original wall finish show through. The walls may have to be painted a light color first. Painting can be avoided by the use of padding or lining.

Padding

There are several other reasons for padding or lining a fabricked room besides that of serving as a color neutralizer. Very rough or cracked walls, tiled or paneled walls need something to smooth out the unevenness. Padding prevents ridges in lath-mounted fabric from showing. It can serve as a temperature or noise insulator. But most of all, padding adds a look of luxury. The elegance of padding is not as evident when used under a busy, dark pattern. But it is shown to great effect when used under a solid color or under a formal fabric such as a shiny taffeta or moiré.

Kinds of Paddings

Cotton flannel or muslin may be used but its thinness puts it into the lining category rather than into the padding. Batting—either a synthetic or cotton—may be used. It definitely will give both a look and feel of softness. The material that is easiest to work with and that has an outstanding appearance is foam sheeting. It tends to cling to the wall and so few fastening staples are needed. Staple depressions do not show through the fabric when foam is used, although they tend to do so with batting. It comes in such wide widths that a lot goes up at one time. It is not expensive if the right source can be located. Foam factories or outlets have this sheeting available in 5-foot-wide rolls in various thicknesses.

Installing Padding

Padding or lining is handled separately from the fabric covering. In the case of backtacking, the fabric is stapled under the upholsterer's tape and the padding is cut to fit, placed over the seam, and stapled into place just before the fabric panel is turned right side out. A padded pillow effect will be created at each seam where the staples and upholsterer's tape press through the foam. If a flatter, tailored look is preferred, the foam must be cut short of the staple line before backtacking. The next row of padding butts against the last row of padding. Foam padding may be glued in place.

Furring Strips

Furring strips are thin pieces of wood that are attached to the wall to be used as a frame for the fabric. Although much has been written about the need for them, actually there are few situations that require their use. The strips are an expense that might be applied instead to the visible finished material. Installing them may be more time

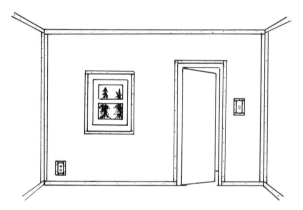

consuming than completing the rest of the job. Once the wood furring system is begun the strips must be installed around the total perimeter of the room—not only ceiling and floor, but also around the openings of doors and windows as well as the vertical spacings where panels are joined. Conditions that warrant installation of furring are: concrete walls (the furring can be glued on); plaster walls that will not accept staples from an available staple gun; walls that are not level, as where the bottom half of a wall is tiled; lease restrictions that forbid holes being put into walls but may permit furring to be glued on. Cracked or uneven walls do not automatically mean furring must be used. They can be covered adequately by choosing a medium or dark background or an allover print, or by padding or lining.

If furring strips are used, place them accurately (based on plumb lines) at the place where the fabric panels will be joined. This will eliminate the need for rechecking plumb lines when the fabric is stapled on. Once the furring strips are up proceed with whatever method of stapling is to be used. Though commonly considered a method, furring strips are a preparation, so they work with either backtacking or flat stapling.

Cornering

Here is the best way to turn a corner. Do not cut the fabric at the corner. Smooth the panel right into the corner, stapling directly through the right side of the fabric and into

Opposite page: There are no corners to cope with in this room. Its architect called it an "elliptical saloon," but it is known today as the Blue Room of the White House. The walls are hung in a striped satin material in two tones of cream. A blue-draped valance trimmed with a tasseled border encircles the room. Copyrighted by White House Historical Association. *Photograph by National Geographic Society*

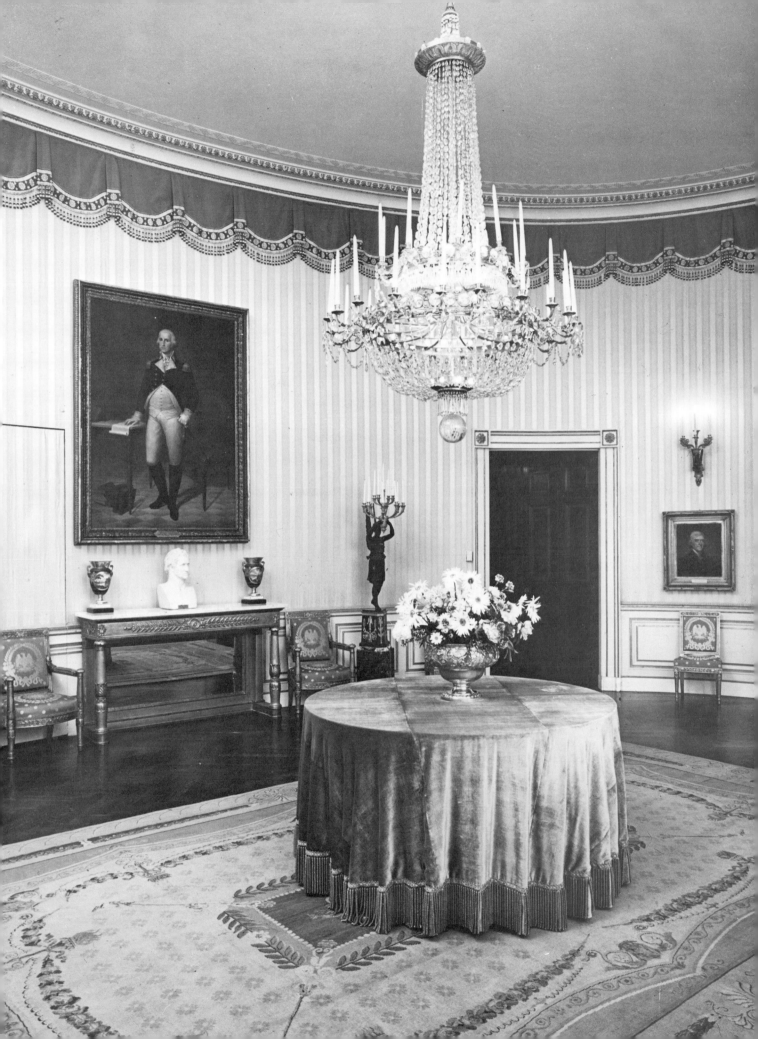

the wall as close to the angle of the corner as the staple gun will allow. Continue the remaining fabric from the panel onto the adjoining wall, stretch it out, and treat it as though it were a separate panel. Be sure to run a plumb line at the end of this panel where the right edge falls. The reason for this is that if the corner is not plumb, the excess fabric can be folded back into the corner and the correction will be barely noticeable. A simple band of self-welting is later glued into the corner to cover the staples and any adjustment of fabric that may have been made.

Window and Door Openings

The professional way to do window and door openings is to work over them as though they didn't exist. Simply staple all around the opening. Then cut the opening out with a single-edge razor blade or utility knife. Handling openings this way eliminates the need to fiddle with pattern matching, a situation where it is easy to get the pattern out of whack.

There are situations where openings are better handled by stopping and starting of panels, for instance when working with a solid color and pattern matching is not a factor. Also, if the panel will end before reaching the middle of the opening, it may be better to end the panel along the near edge.

Spacing

When you use the flat stapling method with molding or trim applied to cover the staples and seams, the spacing of panels becomes important. This is especially true when the fabric is narrow and when there is a great contrast in the finished trim. Think of the fabric panels in the same sense as they would appear used in a traditionally paneled

The placement of the fabric panels lines up with the door and fireplace.

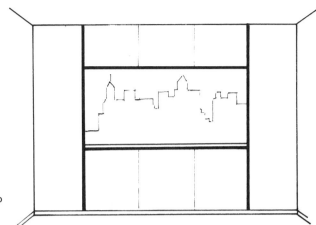

The trim around the window is used to emphasize the view.

room. The result can be architecturally dramatic when trim, door and window openings, pattern repeat, and spacing of the panels all work together. On the other hand, poor spacing can result in a hodgepodge of opposing, distracting lines. Planning ahead is necessary.

Solutions for Problem Areas

Plaster Walls

Plaster, since it is more dense than plaster board or sheet rock, is more difficult to staple into. The densities of plaster vary, however, even within a room. The only way to find out whether a particular plaster will accept staples is to try. Try a variety of staple sizes, too, and, if necessary, try different staple guns. An electric staple gun will work also—even a small, home-workshop type. If the available staple guns can't handle the plaster, use the furring strip system.

Metal Edges

Free-standing walls, corners that jut into a room, and window and door openings that have no wooden moldings framing them usually have a metal strip buried just beneath a thin coat of plaster. The staple gun will not work in metal so these edges must be glued. Hold the fabric in place with a baste staple (it will go far enough in to hold) or with masking tape until the glue dries. It may also be possible to wrap the fabric around past the metal before stopping.

Napped Fabric

If the fabric has a nap, all panels must be placed so that the nap runs in the same direction. Corduroy, velvet, suede cloth, and fake fur are all napped fabrics. Corduroy and velvet should be placed with the nap running up to obtain the greater intensity of color; with the nap running down (it will feel smooth when the hand is brushed over it from ceiling downward) to show off the shaded or frosted effect of the nap. Whichever choice is made, stay with it for every panel.

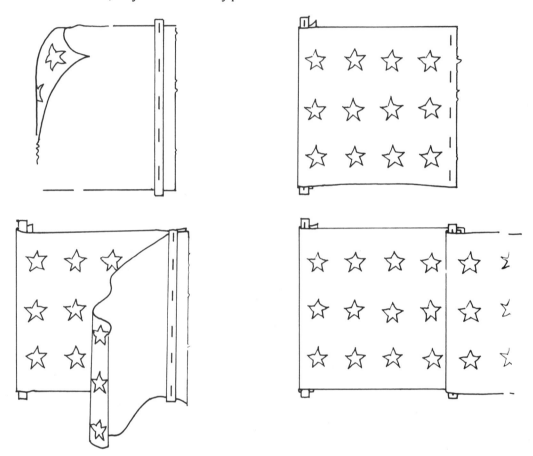

In backtacking the fabric should be stretched tautly lengthwise, but as the panel is turned right side out it should be "floated" widthwise. There should be no break in the pattern where the second panel overlaps.

Backtacking a Complete Room

It is easy to backtack a complete room. No staples will show on any of the vertical seams. Since decorator fabrics are quite wide and sheets can run up to 108 inches, relatively few panels will be required when covering the walls of an average-size room. If they are noticeable, there are easy ways to cover the staples along the ceiling and baseboard.

Some of the information in this section was supplied by Rob Hardy, an interior designer who is the founder of Walls, Uph., which specializes in upholstering walls. Along with a good strong grip, he has developed some innovative techniques.

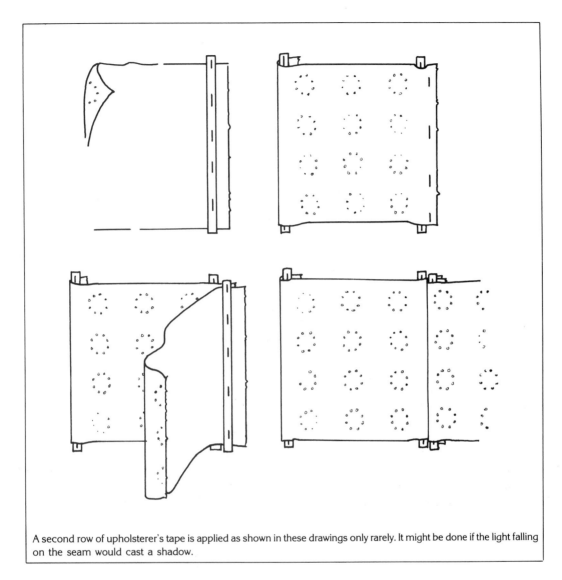

A second row of upholsterer's tape is applied as shown in these drawings only rarely. It might be done if the light falling on the seam would cast a shadow.

Supplies and Tools Fabric, scissors, single-edge razor blades, upholsterer's tape, white glue, staple gun and staples.

Where to Start? Where Will It All End?

Start in a corner of the room. Which corner? The least conspicuous one or the corner nearest to the entrance or the one hidden by a door usually standing ajar. One reason for starting in the least noticeable corner is that if the fabric is patterned, the chances are that there will be a break in the pattern where the last panel meets the first. Also, starting in the hidden corner provides the opportunity to get the feel and experience of working with the material. After hanging the first panel of any given material, an expert is born!

Putting Up the First Panel

When the entire room is to be fabric covered using the backtack technique, the first panel requires a slight variation in stapling.

Leave selvage edges on unless they are very wide.

Cut fabric panels based on their required length and pattern repeat. (At least 2 inches at top and 2 inches at bottom allowance has been added.)

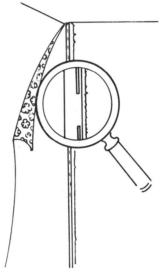

If the last panel will meet the first one, try this trick. Instead of centering the staples on the upholsterer's tape when the first panel is put up, set the staples to the right edge. This will allow just enough room for the raw edge of the last panel to be glued at the corner and then tucked under the first fold.

Staple the first panel as shown in the illustration. Smooth it right side out to the right. Baste staple it in position. Check the right edge with a plumb line. If it is in line, staple the right edge of the panel directly to the wall with widely spaced staples pulling the fabric taut, starting from the middle and working up and then down.

Staple along the ceiling, smoothing the fabric upward as you staple. It is not necessary to fold the fabric into a hem unless the fabric tends to ravel. Trim away the excess fabric at the top with a single-edge razor. Finish the bottom of the panel the same way or, instead of trimming, excess at ceiling and base may be turned under into a hem before stapling.

Place the second panel wrong side up directly over the first panel. Match up pattern repeat. Place a few staples along this edge to hold fabric in place. Place a strip of upholsterer's tape along the edge of the second panel. Staple through tape and both layers of fabric. Place staples centered and closely spaced on tape. Staple entire edge. Turn the panel to the right side and continue in this manner all around the room.

Stopping

One of the joys of backtacking a complete room is to see the raw edge of the last panel disappear behind the edge of the first panel. The final step is to baste staple the last panel into the corner. Trim away all but a half-inch allowance beyond the corner.

Turn this back and run a bead of glue along wall corner. Use the tip of a butter knife or spoon handle to tuck the raw edge under the edge of the first panel. As the first row of staples was set back on the upholsterer's tape for this purpose, there will be just enough room to tuck in the allowance of the final panel. Press down with fingers on glued edge.

As these two photos show, lace and tasseled fringe are two highly ornamental ways to treat the ceiling/wall line. The staples along the top edge are neatly concealed behind the trim. Wooden molding is another good way to finish the edge.

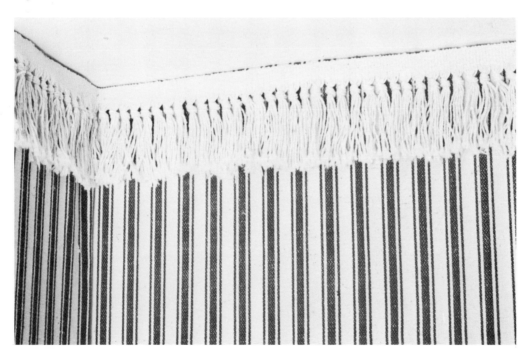

Opposite page: Interior designer Rob Hardy planned these fabric panels to add interest through color and pattern to a formal dining room in Greenwich, Connecticut.

Installed smoothly into the existing wooden molding, a neat finished touch was added by outlining the panels with double welting made from the same fabric.

Below: A tile-patterned fabric is stapled to the walls of this dining room. Since the backtacking technique was used, the joins in the fabric panels are not visible. Room designed by Richard Plouffe. *Photo by Frederick E. Paton*

Pleated Walls

Fabric pleated on walls, by its very abundance, gives a special kind of luxe to a room. Since the recommended ratio of fabric fullness to wall area ranges between 2 to 1 and 3 to 1, the price per yard of the fabric is a consideration. But the elimination of repairing walls in need of resurfacing may result in pleated walls being the more economical method.

There are three kinds of pleats used when stapling fabric to walls.

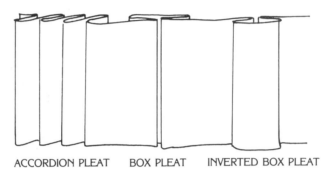

ACCORDION PLEAT   BOX PLEAT   INVERTED BOX PLEAT

If the fabric is a solid color or if it has a random pattern, the pleat repeat is measured and marked before the fabric is stapled up. If the fabric has a design that can be incorporated as the pleating pattern, such as stripes or checks, work directly from the pattern while stapling.

Caution. Pleating a fabric can drastically change the rhythm of the design.

Black, white, and gray checked fabric when pleated can change into a gray and white stripe or a black and gray stripe depending on where the pleats are made.

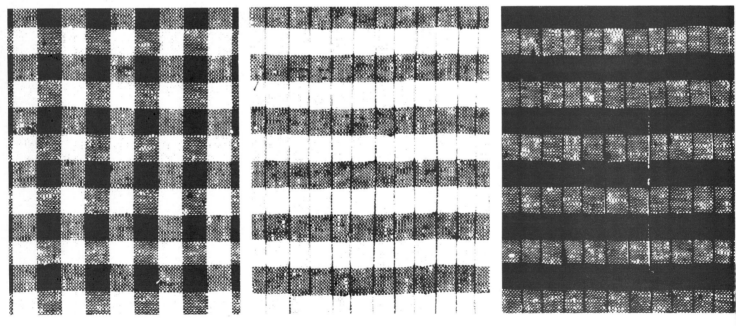

Staple the top edge first when pleating. Stretch fabric downward so that it is under tension when stapled along the bottom. Side seams do not need sewing or backtacking or fastening in any way, as the joins can be hidden in a fold. Stubborn gaps can be shut with a dab of glue. The need to match pattern repeat holds true for pleating also. What can be time consuming about pleating walls is that each opening—window, door, or whatever—must be fastened above and below the break.

The same fabric appears quite different when gathered into draperies and a shirred wainscoting than it does when stretched flat out on the wall. *Courtesy Springmaid Sheets*

A quilt to the right, another to the left, a fragment above the door, and two more set into door panels. Old patchwork on the walls of this entry hall extends a cheerful welcome. Thanks to a combination of layering, backtacking, and cotton bias tape, all staples used to fasten these quilts are hidden.

Opposite page: Shoemakers' children don't go barefoot, and the owners of a fine fabric firm don't have bare walls. Anthony Tortora and Jay Crawford had their glazed aubergine chintz (from their fabric house, Quadrille) custom quilted in a trapunto manner. It was outline stitched into squares and then stuffed from underneath. A practical bonus—the fat quilting, a sort of soft sculpture, proved to be a surprisingly effective sound buffer. *Photo by Richard Champion*

Shirred-Fabric Walls

Shirred walls are similar to pleated walls, the ratio of fabric to wall surface is the same. Shirred walls are in effect pleated walls with random gathers.

Sheets are wonderful instant-shirring panels. Thread a sturdy nylon cord through the small bottom hem of the sheet. Attach as many ironed sheets to the cord in this way as will supply the proper fullness for the wall. Fasten the cord on each end of the wall to a screw or nail. Adjust the sheets along the cord so that the gathers are evenly distributed. Staple through the sheet gathers and the cord into the wall at the ceiling line. This will keep the cord from drooping.*

In most cases the sheets will be longer than the height of the room. They may be left falling loosely on the floor, or the excess hem may be trimmed off and they may be staple shirred onto the wall. Again it is not necessary to seam the vertical panels, as the joins can be hidden in folds.

Quilted Walls

An old Transylvanian saying goes, "Use a heavy-duty staple on an old quilt and you'll feel it pierce your heart," or something to that effect. But the fine-gauge staples in a desk-type stapler are about the same size as the needles originally used to stitch the quilt. A good way to mount a prized quilt is to sew it to a fabric backing—a new sheet will work fine. Let an extra few inches of the backing extend at the top. Staple the backing to a narrow wooden board and roll it under until the quilt lines up along the support. The board can then be mounted on the wall and the quilt will hang safely without wear and tear on it.

---

*If the pattern can't be turned upside down, this quick method can still be used by running the cord through the wider top hem. The gathers will not, however, be as tight and crisp.

To Quilt Fabric on the Wall

Line the wall with batting or foam padding. Staple fabric over it in one of the various methods, but do not stretch the fabric tightly. A bit of give will be needed to form the mounds and depressions. Determine the quilting pattern desired—diamond, square, channel—and mark it with pins. Use decorative brass upholstery nails to sink into the wall. For a more casual look colored thumbtacks may be used. Or with the staple gun shoot two staples in the form of a + right through the fabric, padding, and wall. Cover the + with a glued-on button, knot of yarn, tassel cut from ball fringe, etc. A variation of quilting directly on the wall is using the backtacking method to form floor-to-ceiling channels. In this case the padding is added in a narrow strip between each row of upholsterer's tape. This takes patience and a good deal of accurate measuring. Free form shapes like those used on the piano bench cover on page 110 can be done quickly.

In this bathroom a Roman shade matches the fabric on the wall above the tile. To tie window and wall together, a panel of composition board wrapped in fabric lines the inside of the window recess. It also serves to cover the staples and raw edges of the wall fabric.

In this bed-sitting room, block-patterned fabric is attached to timbered alcoves on either side of the fireplace. Compatible prints are used as throws and hangings on the metal compaign beds. Designed by Richard Plouffe. *Photo by Frederick E. Paton*

Paneling

Another way to treat walls is to staple a covering onto panels that are then fixed in place. With this technique a smooth, finished surface is created with staples, folds, and seams concealed in the back of the panel. This is a technique often used in stores to create smooth, fresh backgrounds in display cases and windows. It is a useful way to cover exposed areas such as archways or the many surfaces around dormer windows. These edges have a very finished appearance with no need for covering raw edges and staples. Quarter-inch panels of compressed cardboard layers work very well, as do foam insulation panels, plywood, or paneling. The panels may be 4 feet wide and ceiling height, or they may be cut in narrow widths to simulate random wood planking. The panels may be left flat or padded with any thickness of padding.

If walls are in very rough shape or uneven from partial tiling, this method evens out the wall surface. Add furring strips around the areas that are not tiled to bring out the surface. On the other hand, if walls must be protected from nail or staple holes, this system can be used with the panels held in place with heavy double-faced tape. This method allows the panels to be taken down and moved to a new home.

A Staple Gun for Paneling

The Whammer gun works like a heavy-duty staple gun except that instead of staples it takes stacks of nails. More detailed instructions come with the gun, but these are included here to show how paneling a room works. Some of these hints apply to other projects in this book.

Panels may be applied directly to walls. For uneven walls, furring strips are required. Apply vertical furring strips every 48 inches (the width of a panel). Remove moldings and baseboard. When removing baseboard, note that the nails are placed at each stud location.

Mark each stud by placing a strip of masking tape on the floor. Panels must be installed with edges lined up over studs.

Proper installation of the first panel is very important. The leading edge must fall on the center of a stud and it must be perfectly plumb.

If the starting corner is irregular or out of plumb, use a divider to scribe the panel for cutting. Cut the length of the panel ½ inch less than floor-to-ceiling height. This will allow a ¼-inch space at floor and ceiling.

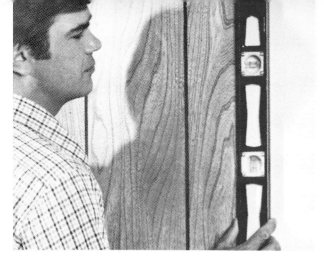

Chalk a plumb line at the starting stud using either plumb or level.

Nails are placed 8 inches apart around all edges of the panels and 12 inches apart on intermediate supports. Start nailing either from the center and work toward the edges, or start at one edge working across to the other edge. Do not nail both edges before nailing the center, as the panel may buckle.

The panel must be cut with back side up when you are using a power saw. For hand saws cut with front up.

To easily mark panel for outlet boxes, tape a piece of carbon paper over box with carbon side out.

Position the panel on the wall, then thump the area of the outlet with the palm. Result will be a perfect impression of the box on the back side of the panel.

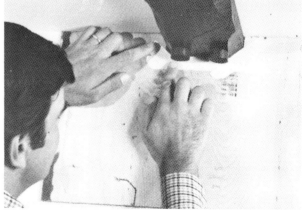

To fit panel around obstructions, make a template on a piece of scrap paneling. Cut it roughly, then use tape to fill in the exact fit. Use this as a pattern when cutting the panel.

Courtesy Swingline Staplers

For what may be one of the few nononagal (comes after octagonal) ceilings east of the Mississippi, a tent look was created. The Indian elephant cotton print is stapled flat to the ceiling in pie-shaped wedges. Tightly shirred, the same fabric on the walls changes the rhythm of the design and adds the additional interest of texture. Designed by Richard Plouffe. *Photo by O. Philip Roedel*

Ceilings

There are many ways to put fabric on a ceiling. One way is to treat it as though it were another wall, stapling the fabric flat to the ceiling. Another way is to tent a ceiling—and there are variations on tented ceilings.

Still another way to treat a ceiling is to create a canopy effect—and again, the variety of ways to do this is endless. But in each case the fabric will combine attaching with suspension or fastening with draping.

Whatever method is used, a fabricked ceiling does not go unnoticed. It belongs to the Drop-Dead school of decorating. It does not create a look of spaciousness, but it

Supergraphic textiles and a staple gun combine to create a unique wall mural and matching window shade. *Courtesy N. Erlanger, Blumgart*

Three toadstools in a woodsy nook make a good spot to read about elves, gnomes, and other middle-earth people.

Laura Ashley fabric covers a table in her sitting room in Chelsea, London. *Photo by Nick Ashley*

Silk, satin, and velvet scraps in a jumble of hot color cushion this piano bench.

A junk shop desk is transformed with staple-gun decorating.

The curved screen is padded and covered in a luscious gooseberry-print chintz. Skirted table, lamp shade, and chair are all covered with fruit and flower designs adapted by Marella Agnelli from nineteenth-century Italian document prints. *Courtesy Abraham-Zumsteg*

A rogue's gallery custom framed with scraps from the remnant basket.

Light filtering through this screen brings to mind stained glass from the Art Nouveau period.

Fabric hinges on these tall linen screens create an interesting crisscross pattern.

Not a stitch of sewing is needed to duplicate this seating module.

A parsons table changed into a strawberry patch with legs.

A sheet stapled on sets this picnic table for a party. *Courtesy Utica for J. P. Stevens*

Fake sheepskin is used to make a cozy desk chair out of a 1940s relic.

Padded and upholstered, this pair started out as inexpensive wooden folding chairs

Interior designer Joseph Widziewicz created sunshine, windows, and curtains for this model room setting with staple gun and fabric. *Courtesy Macy's*

Graphic possibilities are super when working with fabric wall hangings. *Courtesy N. Erlanger, Blumgart*

An Oriental print used to cover a plywood table sets the entire Imari theme for this airy porch by designer Ron Cacciola. *Photo by Bill Rothschild*

Tall, narrow hinged screens are covered in the same attractive red fabric used to cover the walls and table in this dining room. The screens stand before the door to the garden of the New York townhouse of designers Jay Crawford and Anthony Tortora. More screens are used at high windows at the opposite end of the room. *Photo by Richard Champion*

The colorful print of this fabric covers walls and ceilings and all the angles and levels between the two, tying the room together and transforming it into a flower bower. Chair and window settee are upholstered in the same flowered fabric from the fabric house Quadrille.

does give a feeling of intimacy and coziness. Whatever method is used, one thing must be kept in mind—the law of gravity.

Stapling Fabric Flat on the Ceiling

Professionals prefer to machine stitch enough panels together so that the entire ceiling can be put up in one piece. If there is a pattern, the fabric panels are stitched so that the design will be exactly centered—lengthwise as well as widthwise. But a beginner will find it easier to staple the fabric up in separate widths.

If there is no pattern, or if the pattern is an allover one or striped, start stapling one

selvage edge along the longest wall. Do not attempt to staple on the ceiling (it is awkward to use the staple gun upside down), but along the wall as close to the ceiling as possible. If the walls are to be covered in fabric, the staples used to attach the ceiling fabric will be covered. If only the ceiling is to be done, they can be covered by welting, molding, or braid.

Mitered Ceiling

The directions just given are for putting up fabric in panels that run lengthwise from one end of the room to the other. Another way is to use the fabric so that it radiates from a center point out to the corners. Striped fabric is often used, formed into a mitered design of four triangular pieces. (Some stripes are not symmetrical, and so will not give a perfect match as the one in the photograph does.) If possible, clear the room of furniture and lay out the fabric on the floor.

Fasten two strings to the ceiling, running diagonally to the four corners. Leave strings in place and use them as guidelines. The strings cross in the dead center of the room. The fabric may be stapled up in separate panels or seamed into one piece before being installed.

Sunburst Ceiling

The sunburst draping requires a great deal of fabric.

The center of the ceiling is located by attaching strings from the corners. A block of wood (6 inches square × ¾ inch thick is a general size) is fastened securely to the ceiling. Be sure that it is anchored in a beam or some wooden support. The fabric is then stapled in tight, overlapping, random pleats onto the edges of the wood block. After the fabric for one side is stapled to the block it is stretched out to the wall and stapled in random pleats. The same is done for each of the other three sides. Traditionally a large rosette is made out of the fabric and fastened in the center

An alternate way is to gather the fabric into a tight cluster, wrap it with nylon cord on the wrong side, and tie this to a long molly screw fastened in the ceiling. The fabric is stretched out to the walls and stapled. The yardage can be worked out so that the fabric is either flat or gathered when it reaches the wall. When estimating yardage for either the mitered or sunburst style, remember that the panels running on the diagonal from corner to center must be longer than those running from the middle point of the wall to the center. Allow for this, both when estimating yardage and when cutting panels.

Opposite page: This green-and-white-checked bedroom was one of the most copied rooms of the 1960s. It was designed by Bert Wayne and John Doktor for Anne Klein, the late fashion designer. The checks cover ceilings and walls in a flat, tailored manner. The crystal chandelier shares the room eclectically but companionably with the $1-a-yard gingham. *Courtesy Wayne and Doktor Ltd*

Following page on left: The stripes on this mitered ceiling match up perfectly. A wall-to-wall cornice has been padded and covered in the same striped fabric. Dining room designed by Jane Victor of Manhattan.

Following page on right: This dining room gazebo is heavily draped in the optimim 3 to 1 fullness ratio—and then some. The sunburst ceiling has a wide valance surrounding it. The tulip-print fabric, made of Celanese acetate, is used flat against the walls. Over this traditional draperies are hung and loosely caught back. The entranceway is draped, too. Designed by Shirley Regendahl.

5 Things to Sit On

The fine craft of upholstery developed with the search for a comfortable place to sit. It was based on a mysterious inner build-up of many materials—metal strappings and springs, woven webbings, jute twine, horsehair padding, kapok, feathers, and down—all based on a balance of resistance and give. Most materials were organic, and the final covering hiding these ingredients most likely was a sturdy mohair cloth. And then along came materials new to the furniture world—foam rubber (replaced now by rubberlike synthetic foams), poly pellets, and synthetic battings.

Simplified furniture goes with our simplified life-style. Today the upholsterer who works in the Old World traditions has the role of craftsman of a disappearing art. With the furniture assembly line came the use of the staple gun. And there is no need to apologize for upholstering with a staple gun — staple gun upholstery is not a tacky thing. Most furniture makers use automated staple guns (automatic glue guns, too), but a simple hand-operated one can do a fine job for the home user. While a heavy-duty gun might be needed in some cases—depending on the hardness of the frame and the type of covering used—it is not always necessary. Tacks can be used to supplement staples in places where a lot of strain will occur.

The advantage of working with a stapler—rather than tacks and hammer—is that one hand is free to control the fabric while the other is setting staples with dead accuracy. Also, staples put a smaller hole in the fabric and, because of their narrow gauge, take up less space in the wood frame. This can be a big factor. For instance, if a muslin undercover is used, it must be completely tacked along all edges. On top of this comes the outside covering, again tacked down along all edges, and finally gimp or welting needs to be applied to cover the previous tack heads.

This book does not attempt to give a course in upholstery, just to take away some of the mystery surrounding it and to show ways that a staple gun can give you new things to sit on.

The projects in this chapter touch on three basic areas—re-covering existing objects, redesigning old pieces, and creating new ones.

Opposite page: Interior designer Ron Cacciola created this welcoming sitting room using a mix of Alan Campbell fabrics. The printed canvas on the chaise and its companion on the wall behind are stylized interpretations of the bamboo motif. A peaceful coexistence with fruit, flowers, and baskets is evident. The chaise, constructed from a shaped plywood box and foam bolster, was padded with batting and foam and then upholstered with a single length of fabric. *Photo by Patricia Lambert*

The Hoffman Chair. Two virtues of this chair are its good design and the fact that it is a stack chair. A third marvelous feature that is not detectable is the fact that its designer, Pat Hoffman, planned it to be easily re-covered by an amateur using a staple gun. It comes with a padded, wool-covered seat that is secured to the molded hardwood frame by two screws. *Courtesy International Contract Furnishings Incorporated*

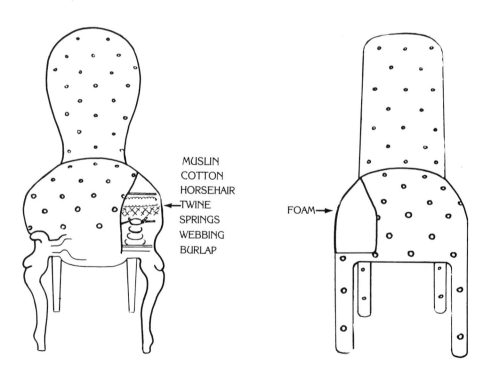

MUSLIN
COTTON
HORSEHAIR
TWINE
SPRINGS
WEBBING
BURLAP

FOAM→

Slip Seat

A slip seat is a removable wooden seat that fits into the framework of a chair. Re-covering it is a simple staple gun project. Almost as simple is adding a slip seat to a chair that has none. An upholstered seat can give a whole new look to a wooden chair, and a chair in need of repair—perhaps a caned seat has worn through—can be salvaged by cutting a new seat from a piece of plywood. Stapling a seat cover costs a fraction of the price of tie-on cushions and gives an unlimited choice of color and materials. Suede, quilted cotton, a different bright color for each chair around a dining table, a mix and match of small provincial prints—the possibilities are endless.

If a delicate fabric, a pale color, or a fragile old quilt is to be used on a dining chair, the slip seat may be protected with a layer of lightweight clear vinyl. Unlike ugly plastic slipcovers, if smoothly stretched and stapled to the bottom of the seat, the fabric will look as though it has been glazed.

Remove the slip seat from the chair. It may be held in place by screws on the underside of the frame. Strip off the old covering. If the chair is very old, the padding may be cotton batting that has become flat or foam rubber that has begun to disintegrate. If so, replace with a layer of foam or batting or both. Besides making the sitting softer, a padding gives the seat a smooth, finished appearance. Lay the stripped seat down on the new padding and trace the outline. Leave enough of an allowance so that the padding will turn down over the top edge of the seat, but not so much that it will

Staple in the center of one side and then in the center of the opposite side, keeping an even allowance on both sides so the pattern will stay slightly centered. Do the same on the other two sides, pulling fabric with one hand while stapling with the other. Notice in the photograph that the thin layer of foam padding extends around the sides of the seat, but not so much as to interfere with stapling the fabric firmly onto the seat.

Work around the seat, pulling and stapling opposite sides. Avoid folding or pleating the fabric, as this could prevent the seat from fitting back into the frame. Stretch the wrinkles out by pulling the fabric toward the center of the seat.

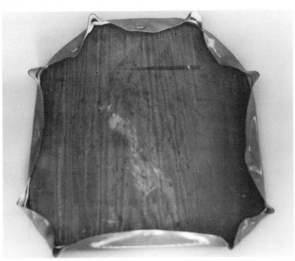

Continue stretching and stapling until the distances between staples fill in. Even though there are still small puckers between staples, the side edge of the seat is smooth.

When the fabric is securely stapled, trim away the excess with scissors. Return slip seat to frame and replace screws.

prevent the seat from fitting back into the frame. If the new covering has a design, lay it right side up on a smooth work surface and center the slip seat on it to be sure that the pattern is properly centered. With chalk or pencil draw around the seat, adding 2 inches on all sides. (This amount will be right for plywood up to ½ inch thick and ½-inch foam padding.) Cut the fabric and flop it over so that it is facing wrong side up. Place the padding on it and the seat on top of the padding.

If the seat is a new addition, there are two ways it can be fastened on. If the covering is woven or knit fabric (nails leave holes in plastic), small brads may be hammered in at an angle through the side of the seat and into the frame. With the point of a pin, raise the surrounding threads of the fabric over the head of the brad and close over the separation made where the nail entered. Another way to secure the seat is to screw from underneath through the frame and partway through the new seat.

Supplies and Tools Chair with a removable slip seat, fabric to cover seat, padding if desired (foam or batting), staple gun and staples, scissors.

These Mansfield Manor chairs are an excellent sample of what's new in contemporary chairs. As the trend turns toward upholstered walls, tables, and beds, completely upholstered side chairs are very evident, too. The straightforward, appealing shape of these chairs is complemented by Abraham-Zumsteg's Winterthur fabric.

Covered in gold satin, this hassock was given a matching tassel and renamed a gout stool.

Stool

This hassock is one of the easiest projects in the book—it can be started and finished in about an hour. This tube came from a carpet store and was cut to size with an electric hand saw. It can be cut to any desired height and padded little or lots. The choice for coverings is unlimited.

Supplies and Tools Heavy cardboard tube (this one is 15 inches in diameter, 9½ inches high), batting, one square yard of fabric, marker, scissors, corrugated cardboard, staple gun and staples.

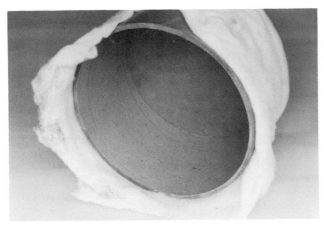

With a utility knife, cut a circle from composition board to fit the top of the tube. Glue it on with white glue. When dry, wrap the top and sides in batting.

Fold the fabric into quarters and round off the corners with a scissors so that the material is now circular. Cut a notch through all four layers at one edge so that circle is marked into quarters. With a marker, divide the tube into four parts.

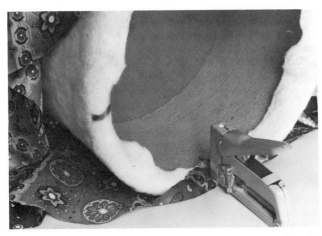

Place the fabric wrong side up on the work surface and place the tube top side down on it. Center tube so that fabric notches line up with marker points. This will help distribute the fabric. Staple fabric to the inside of the tube at the quarter marks, smoothing fabric tightly. Stretch fabric with one hand while stapling with the other, alternating around all sides of the tube, continuing to distribute the gathers evenly.

Trace the new hassock on corrugated cardboard or composition board, and cut out circle. Wrap circle with fabric and staple it on. Use white glue to secure bottom to hassock. Let dry.

When fabric is completely stapled, trim excess with scissors. Attach tassel by punching hole in center of top with nail. Poke cord through to inside and tie it around a small wooden dowel or popsicle stick.

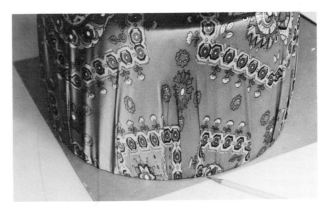

Three toadstools in a woodsy nook make a good spot to read about elves, gnomes, and other Middle Earth people. The stools are not difficult to make.

Toadstools

First locate a log. If none is lying around the property, try a tree surgeon, power company road crew, or a sawmill. It is best to use a dry log so the bark will chip off easily. Sometimes it can be peeled away by hand. Fill any large cracks with plastic wood filler available in small cans from hardware or paint stores. The stripped log may be left as is for a woodsy, natural look or sanded and given a coat of sealer. The sanding can be time consuming, but developing the satiny patina of the wood is a satisfying pastime.

Supplies and Tools (The three toadstools were made using molded foam pillows in 14-, 16-, and 20-inch sizes. The figures given in the directions are for the 16-inch pillow. The dimensions for the other two sizes are listed separately below.) Log about 12 to 14 inches high, 16-inch foam pillow, wood filler, ¾ yard of fabric, two circles 11½ inches in diameter cut from ¼ or ⅜ inch plywood, white glue, felt, screws, screwdriver, scissors, drill, staple gun and staples.

For 14-inch pillow One 10-inch log, two plywood circles 9½-inch diameter cut from ¼ or ⅜ inch plywood, two circles 22½-inch diameter cut from ⅝ yard of fabric.

For 20-inch pillow One 18-inch log, two plywood circles 15½-inch diameter cut from two circles 31½-inch diameter cut from ⅞ yard of fabric.

To measure for the 11½-inch plywood circles, tie a pencil on a piece of string and use it like a compass. Saw out two circles.

Use the same pencil and string method to mark circles on cloth. Velveteen was chosen for these covers for its likeness to the velvety texture of the real thing. Cut two circles from the fabric, one 27 inches in diameter, another 13 inches.

Place the smaller fabric circle on the work table with the wrong side facing up, and center one of the plywood circles on it. Pull the fabric onto the plywood and begin to staple. Fabric can be distributed more evenly if the first four staples are placed in the form of a cross in line with the vertical and horizontal weave of fabric, rather than on the bias. The fabric should be tightly stretched without folds or pleats.

Midway between each staple, use one hand to pull fabric tightly toward the center of the circle and staple there. Continue to staple, each time centering the staple between those to either side and working around the circle alternating sides, so that the fabric is taut and evenly distributed.

Place the larger fabric circle wrong side facing up on the work table and center the 16-inch foam cushion on it. Place the uncovered 11½-inch plywood circle on the cushion. Make sure the three are centered. Begin stapling the fabric up around the cushion and onto the plywood circle. Alternate sides while spacing staples evenly around the circumference. The fabric will form many tucks and pleats.

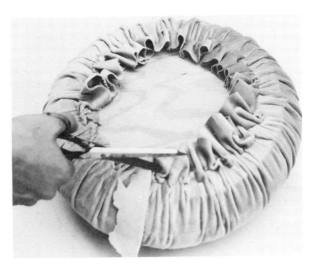

After the fabric is firmly attached to plywood, use a scissors to trim away excess.

Center the fabric-covered plywood circle on the top of the log. Attach it firmly to the log with four screws. An electric drill makes this a quick job.

Place the log upside down on the fabric-covered cushion. Center the log and cushion so that the two circles set evenly on top of each other. Screw the two circles together. Using white glue, attach a piece of felt to the base of the log to prevent its scratching floor or snagging rug. Turn right side up.

Ordinary wooden beach chairs serve as the inspiration for the window display at Clarence House. But they are not covered in the usual striped canvas. Robin Roberts covered them with precious real silk striped in hot sunshine colors. The idea can be copied with any fabric. For strength fold a layer of denim or canvas inside the fabric before stapling it on. *Courtesy Views & Reviews*

Fuzzy Chair

The appearance of the subject matters not at all. In fact, a shabby-looking one will increase the pleasure of transformation. But the chair should be sturdy in order to avoid joint glueing or reinforcing.

A cozy desk chair was made from a relic of the 1940s. Fake sheepskin was used to cover it, but the same method works with any fabric.

The first step is to fill in open areas in the chair frame. Thin plywood or inexpensive wood paneling works well for this project. Composition board or even heavy corrugated cardboard may be used, provided the fabric will not be stapled to them without a wood support underneath. Wood strips can be glued into place to reinforce edges that anchor the upholstery.

If the back or legs have any knobby extensions, saw them off so that the chair will be as squared-off as possible and so that the plywood will fit snugly against the frame. Using the chair as a pattern, place it on the plywood and trace around these four sections—top, back, bottom back, and two side bottoms. Leave at least one inch of free space between top and bottom back panels so that the seat upholstery can be pulled through from the front and secured on the outside back seat frame. If the back of the chair is quite straight a single panel may be used, but a slot extending from post to post must be cut into it at seat level. (The back will be covered last with a single piece of cover fabric.) Attach the panels to the chair with screws; glue them on if made from cardboard. Cut a new pad for the seat from foam, adding a ½-inch allowance to the sides and front. Glue seat to chair with foam cement. Add a layer of batting on top of the foam large enough to extend around the sides of the seat.

Measure the length and width of the seat and add 6 inches to each measurement. Cut the seat piece from the covering fabric. Place the fabric right side up over the batting and center it. Holding it in place with a few basting staples, turn the chair upside down and staple fabric to the front inside edge of the chair. Work from the center out to the ends, stretching fabric while stapling. Turn the chair right side up and smooth fabric toward the back of the chair seat. Tuck it through the opening in the back. With a scissors, cut into the fabric up to the posts so that it wraps around them as snugly as possible, then clip wherever necessary to get a tight fit. Sides are stapled next. Begin at the centers and alternate stapling on either side so that the fabric will not be stretched out of position. At front corners fold the fabric into a pleat. For a professional look, place the fold exactly at the corner and adjust if necessary so that both corners appear identical.

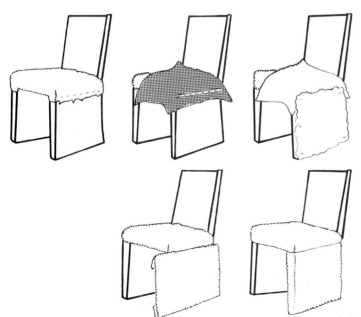

Measure side and cut a piece of fabric to fit, allowing ample extra allowance for the slant that most chairs have. Also, allow extra fabric to wrap around the front and extend to the inside of the legs. Hold fabric in place with right side out to determine correct placement. The vertical grain of the fabric should be plumb and the horizontal grain should be level with the floor, regardless of the slant of the chair.

When this is established, carefully flop the fabric back onto the top of the chair. Staple a piece of upholsterer's tape through all layers to the seat frame. Check by tugging a bit to be sure staples are long enough to stay in. Place a layer of batting over the plywood panel and staple to the back of the leg. Fold the fabric down over the side and staple along inside bottom. Fold front edge and wrap to inside of leg and staple. Back edge gets wrapped and stapled flat along back. Cut fabric for front top with enough allowance to wrap around sides and to tuck through space between seat and chair back. Pad front top with batting cut large enough to wrap over edges and onto back of chair. Place fabric for top in position and staple across top back, starting from center and stapling out toward the posts. Smooth fabric down taut and stuff through slot to back. With scissors, cut into post corners so that fabric will surround posts. Staple fabric to the back of the chair. Now bring the sides around to the back and staple them. Make tailored pleats at the top corners.

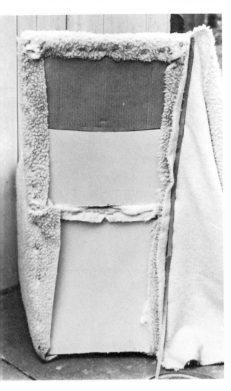 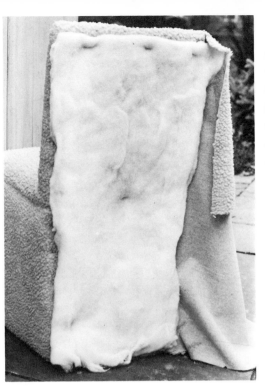 

Cut a piece of fabric to cover back of chair from top to floor plus 2-inch allowance on all sides. Be sure that the grain of fabric remains plumb rather than following the curve of the chair. Using upholsterer's tape, backtack fabric to side of chair. Staple securely into frame, starting in the middle.

Pad back of chair with batting stapled in place.

Bring fabric around over batting and baste staple it in place. This chair was finished by hand stitching with a curved upholstery needle. Tack stripping, shown in the photo on page 12, may be used instead to finish the side edge.

Inside bottom of chair is finished by fabric panels backtacked at front and flat stapled to inside legs at back. A final back inside section is hand sewn. Inside is difficult to work in. If fabric is appropriate it can be glued on, or panels cut to fit the inside can be wrapped and attached with brads.

Robin Roberts has a reputation for creating sensational windows for his Clarence House showroom in New York. He once spent $10,000 on a single display — not the one shown in this photo. With true throwaway chic, he uses white canvas to cushion a wooden porch swing. The idea is free to copy. *Courtesy Retail Reporting Bureau*

Upholstered Folding Chairs

X-ray vision is required to see that these padded and upholstered chairs started out as ordinary wooden folding chairs. The project can be duplicated by anyone with a staple gun and patience.

They were covered in a bright orange felt, durable enough for these chairs, which are folded up and stored in a closet between parties. For a chair expected to get a lot of wear choose a sturdier covering.

Supplies and Tools 1¼ yards of 54-inch-wide fabric, wooden folding chair, foam and poly padding, marker, scissors, glue, pins, upholstery needle, strong thread, four furniture glides, staple gun and staples.

Two small changes are made to the frame. Saw off the top post extensions. Cover the seat with a piece of thin plywood or layered cardboard to straighten the line across the back. Both of these adjustments make the covering process easier.

Place the chair on its side on a sheet of opened newspaper to make a pattern for the post and leg section. With marker and ruler, mark off obstructions caused by bolts or bars.

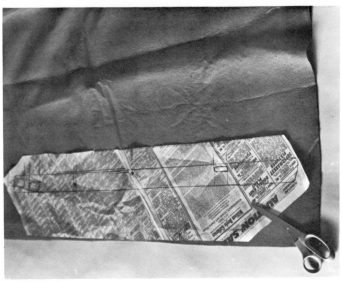

Lay the pattern on fabric. Stretch the fabric slightly by securing it to work surface with masking tape. This insures a taut fit on the finished covering. Transfer the bolt and bar obstructions from the pattern onto the fabric with chalk.

Cut out the leg post piece and lay it on the chair as shown. With scissors, snip adjustments until it fits well.

Reposition the fabric so that it is ready to wrap. Apply white glue to inside leg, spreading it thinly and evenly. (Spray adhesive works well too, if care is taken to mask all other surfaces.) If the fitting around bolts left a gap, cut a piece of fabric in the shape of a small doughnut, make a slit in it, and glue it in place over the gap.

After glue has dried, wrap the fabric onto the back surface of the leg and staple it. Trim off fabric so that it is flush with the edge.

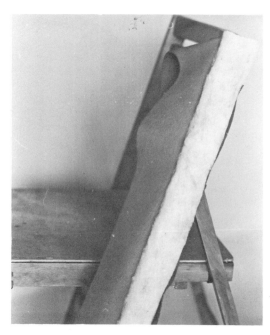

Cut a layer of batting 10 × 34 inches, fold it in half lengthwise. Use it to pad the front and outside of the leg post.

Carefully, so as not to dislodge the batting, wrap fabric around to back of post. Trim off excess fabric so that turnunder will be a consistent width no wider than the width of the post.

Fold and pleat top and bottom of leg post, covering wood completely. Pin fabric closed from top to bottom, making necessary adjustments. This edge can be finished by hand stitching or by glueing. If it is glued, unpin a small section at a time, glue with white glue, and reinsert pins at right angle as in photo to keep tension on fabric until set. Repeat these steps for the other leg post. Use the same technique and steps to cover the front leg posts.

Above:
Cut a rectangle of fabric 16 × 5 inches for the front rung and one 18 × 5 inches for the back rung. Starting from center and working out to the ends, staple these pieces to the undersides of the respective rungs.

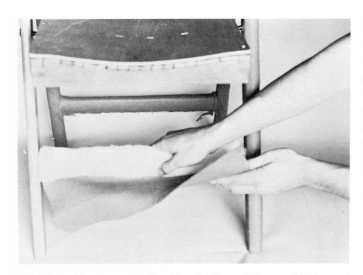

Pad lightly with a layer of batting. Wrap fabric carefully around batting and staple on bottom of rung.

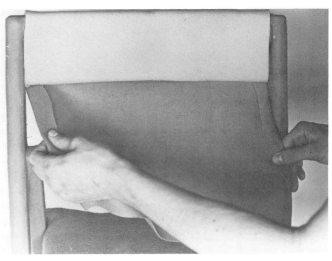

Pad the backrest with a layer of foam or batting. Staple a piece of fabric 18½ × 12 inches to the lower edge of backrest. Wrap fabric carefully around padding and fasten with pins at lower edge, covering original row of staples. Staple, hand stitch, or glue this edge.

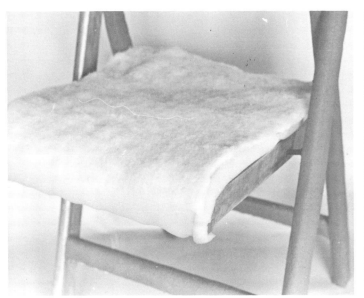

Pad the seat with a piece of 2-inch foam 24 x 24 inches. Place a layer of batting over foam. Cut a piece of fabric 20 x 22 inches. Center this over foam and batting.

Turn chair seat upside down onto work surface and staple seat cover on. Staple front edge first, then back, then sides. At corners fold excess fabric into a tailored pleat lined up right at corner edge. Remove all pins. Press with steam iron to eliminate pin holes and puckers. Close any gaps that may exist with hand stitching. Hammer furniture glides on bottom of legs to protect fabric.

Scraps of velvet, silk, and satin in a bright jumble of color are the main ingredient of this piano bench cushion. The technique used to make it can be described as backtacking on a horizontal plane or as channel quilting. It is a simple way to create an attractive, comfortable seat top.

Piano Bench

Supplies and Tools Composition board or plywood pad cut to fit the bench top, scraps of fabric in assorted widths but at least two inches longer than the pad's width, batting, upholsterer's tape, scissors, staple gun and staples.

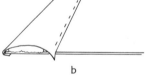
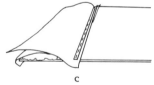

a   b   c

Staple strip of fabric to underside of pad along left edge. Turn pad right side up and place a piece of batting along the left edge. With one hand holding batting in place, fold over fabric strip and fasten it with a few staples. Place a second fabric strip wrong side up over the edge just stapled. Fasten it with a few staples. Place a piece of upholsterer's tape over it and staple through both fabric layers into pad. The strips are cut in random widths and assorted wedge shapes. Besides creating an interesting design, it is much easier than trying to form evenly spaced channels.

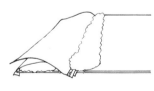

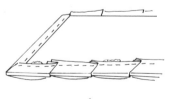

d   e   f

Place a layer of batting snugly against the upholsterer's tape and, holding it in place, fold the second fabric strip over it. Staple the right edge of the second strip in place. Continue adding strips and batting until the entire pad is covered. Turn pad wrong side up and flat staple the last strip along the edge. Turn the fabric over the length of the pad and flat staple at each backtacked seam. Fill in with staples between seams. Do the same on the opposite side. Attach pad to bench with double-faced tape or, if the bench is not fine wood, hammer small brads through the top of the pad and into the bench. Pull fabric over and around the brad heads to hide them.

Seating Module

Make one for your pet's birthday. Make two modules and place them side by side for a cozy love seat. Cluster four in the center of the room for a dramatic seating island. Bank the modules around a room wall to wall to form a banquette. The combinations are endless. Experiment, moving them about until they are in the perfect arrangement. Then wait until a change of mood or a roomful of guests inspires a more perfect arrangement. No other kind of furniture provides this freedom and flexibility for creating an immediate environment.

The design of this handsome seating module has been worked out especially for staple gun construction. The need for a sewing machine or hand sewing is completely eliminated. Although it has the proportions and the finished detail of the expensive modular units pictured in design magazines, it is not a difficult project.

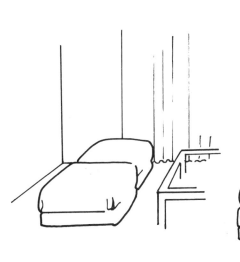

If only one module is made, there will be fabric left over from the 3⅛ yards specified in the supplies list. It is possible to buy less fabric and cover the frame in four pieces rather than one long one. The fabric can be backtacked at each corner. For simplicity's sake, the directions given here use the one-piece method. Consider the leftover fabric a bonus to be used for throw pillows.

If a number of modules will be made, work out a layout plan for the total pieces of fabric required and determine yardage requirements based on it. In general, the more modules made, the more economical will be the use of the fabric. To make this project following the dimensions given, it is necessary that the fabric used be at least 54 inches wide. Measure it before buying to make sure.

The frame, made with simple butt joints, requires two 27-inch lengths and two 25½-inch lengths of lumber. If you don't own a saw, the lumberyard will cut them for a small charge. Use glue and three nails at each joint.

Wrap a 7-inch strip of batting around frame and pat it into place. Secure with a few staples, if necessary. The wood may be rough enough to hold it.

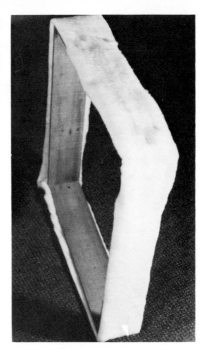

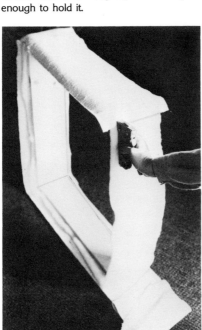

From fabric, cut the following three pieces: 47 inches × 47 inches; 29 inches × 29 inches; 7 inches × 110 inches. Starting one inch below nearest corner, staple 7-inch-wide piece of fabric through batting into frame. Wrap fabric tightly around the four sides of frame.

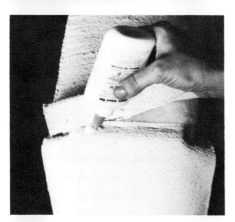

At corner, turn under one-inch hem. If fabric is loosely woven, it may have stretched as it was wrapped around frame. If it has, cut off excess so that one inch only remains to be turned under. Run glue along fabric and press firmly in place.

Holding the staple gun at an angle to the surface rather than flat against it as it is normally held, insert several baste staples to hold the fabric in place while the glue dries. As these staples will be somewhat raised, they can be easily removed later.

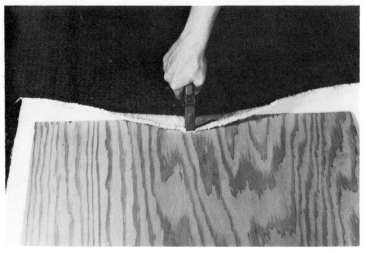

← Wrap fabric around to inside of frame. Staple selvage edge to four bottom sides of frame. Stretch fabric smoothly and tautly over sides and staple to the inside of the four top edges.

Above:
Place 27-inch square of plywood on wrong side of 29-inch square of fabric. Center it. Starting with side that has the selvage edge, staple it in the center. Stretch fabric out to corner and staple it. Fill in evenly with staples between center and both corners. Fold fabric at corners into a flat fold.

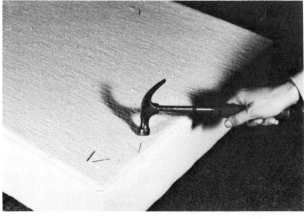

Reverse panel so that opposite side is facing you. Repeat the stapling process on this side, making sure that fabric is tightly stretched. Repeat stapling process with third and fourth sides.

Place covered plywood panel on top of platform and nail in place. Just before giving nail its final blow, use a pin to separate the threads of fabric from around nail head. Force material around and over nail so that it will be hidden.

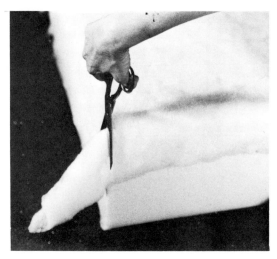

Place a square yard of batting over foam cushion. Pat into place. Cut off extra batting at corners with scissors.

Place foam and batting upside down on wrong side of 47-inch square of fabric. Measure sides to make certain it is centered. Place 26-inch square of plywood on top of foam cushion.

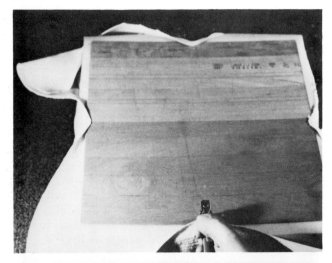

Bring fabric up to plywood and staple it in center of each side. Start with the side that has selvage edge. Staple from center using free hand to stretch fabric toward corner. As you near corner, carefully fold fabric into a pleat, making sure that fabric is evenly divided between this pleat and the one that will meet it on the other side of the corner. Rotate cushion so that opposite side is now facing you. Repeat process of stapling from center toward corner and pleat fabric. In the same way, staple fabric to two remaining sides.

Trim away excess fabric. Place cushion on top of covered frame and secure with a few finishing nails angled through the fabric into the plywood base of the upper cushion and down into the fabric-covered plywood of the bottom frame.

Supplies and Tools Foam cushion 27 inches by 27 inches by 9 inches (two layers of 4½-inch foam or three layers of 3-inch foam may be used). 3⅛ yard of 54-inch-wide fabric, a 27×27-inch frame constructed of 1 × 6 fir or pine board (actual lumber dimensions are ¾ × 5½), 26 inch square and 27 inch square cut from ⅜-inch plywood, batting, white glue, foam glue, staple gun and staples, scissors, yardstick.

6 Things to Lie On

Beds—because of their size and the amount of time spent in them—deserve to look attractive as well as to be comfortable. The sheet manufacturers, with their brilliant designs, have caused a kind of revolution in the look of the well-made bed. (So much so that it is no longer an insult to identify someone as looking like an unmade bed.) Although standard tickings are too often limited to dull gray designs, it is simple to upholster a box spring in attractive new covering.

Wrapping a Box Spring

The easy way to do it is to wrap it in fabric stapled to the underside. What material to use? An obvious choice is a sheet—designwise and sizewise it fits. (A fitted sheet doesn't fit smoothly on a box spring, but a flat sheet stapled on fits like a glove.)

There are other materials to use. Fabric by the yard can be used with striking effect if a room has other uses besides that of bedroom. Woolen suiting, especially gray flannel, is good when a boudoir look is not wanted. A quilted comforter or spread also gives the bed base a smart look.

The average-size single box spring requires fabric 52 inches wide. (The thickness of the average box spring is 6½ inches, width is 39 inches.) Most home furnishings yardage runs in the 50- to 60-inch range, but measure both the box springs and the fabric before buying. Fashion fabrics may be used—woolens are at least 54 inches wide, but cottons and other types will have to be seamed to be wide enough. For a 30-inch-wide day bed the possibilities are endless.

Wrapping a box spring will most likely have to be done on the floor. A hard surface is best. If a carpeted area must be used, make sure that the covering is stretched smooth underneath and stapled tautly.

Opposite page: Bedrooms that serve a dual purpose are not a twentieth-century development. Charles VI (that's the king sitting on his bed) is seen holding a staff meeting. His canopied bed, hangings, and draperies are all done in the same fleur-de-lys pattern, apparently all the rage in the 1400s. From "Pierre Salmon," courtesy Bibliothèque Publique et Universitaire, Musée Rath, Geneva, Switzerland. *Photo by François Martin*

117

Nearly five centuries later, this bedroom has been given another royal treatment. The ensemble look is achieved by using one small repeating pattern on canopy, walls, ceilings, at the window, and on a small corner screen. Quadrille Fabrics describes it as a soft geometric. It serves as a perfect foil for the Amish quilt on the bed.

A clutch of exotic bird-of-paradise in a porcelain vase reflected in floor-to-ceiling mirrored panels. It takes a singular talent to know when the different drumbeat is the one to march to—in this case a simple bed base and mattress casually wrapped in stripes. The base is upholstered in woven black cotton. Giant round bolsters and pillows make it a comfortable lounge. From the Crawford/Tortora design team. *Photo by Richard Champion*

Smooth the fabric out on the floor wrong side up. Center the box spring on it wrong side up. If the box has removable legs, unscrew them. Make a chalk line indicating the location of the screw holes so that they can be easily relocated. Staple along the length of one side, starting from the center and stretching the fabric taut toward each end. Pull the fabric tightly around and up on the opposite side and staple it the same way. Staple the ends, leaving all four corners free. At the corners tuck the extra fabric into a pleat with the folded edge lined up directly at the corner of the box.

Upholstered legs add a classy finishing touch. If the legs are ordinary posts, pad them thickly with batting or foam before covering them in fabric. (Saves on stubbing toes in the dark.) Fat, round legs called bun feet can be bought in specialty hardware or upholstery shops. They can be left plain or wrapped in a gathered circle of fabric. They will lower the height of the bed, but give it a different, lounge look. Another possibility is to leave the legs off and set the box spring directly on the floor.

This bed is set into a very low wooden frame carpeted in a contrasting color to the floor carpeting. A third color covers a ledge between bed and wall. Plywood bases and remnants from the rug store are all that are needed to duplicate this setting. *Courtesy Burlington Mills*

This low bed was built to fit under the sloping ceiling of a musician's loft. The bed platform, made of plywood, has a large table attached at one corner. The whole base is padded, then upholstered with sheets. *Courtesy Martex*

Opposite page: Besides wrapping there are other ways to dress a bed base. "The Well-made Undressed Bed. Or, Look Ma, No More Hospital Corners." The puffy comforter can be folded in half and the pillows plumped in a jiffy. The box spring is covered with a matching full-size sheet, folded and wrapped to the back before the mattress is placed on it. Marimeko patterns for Dan River.

A higher base raises this bed to average height without the need for legs. The plywood base is padded with batting before sheets are stapled on. Another example of the refreshing "unmade bed" look. *Courtesy Burlington Mills*

Hangings for Box Springs

For a feminine look, sheeting is gathered into a very full dust ruffle. If the bed base is plywood, backtack the gathers onto the top. Set the sheets far enough in so that there will be no gap between ruffle and mattress. *Courtesy Burlington Mills*

A box-pleated bed hanging can be attached inside the frame of the bed. Sheets are used for the corner hangings at the bedposts and along the stretchers in between. For an overhead frame like this, use the backtacking method from the inside. The drapery falling over the staples will conceal them. These sheets are designed by Suzanne Pleshette for Stevens Utica. *Fine Arts*

Headboards

There are two types of headboards—those attached to the bed and those attached to the wall. There is no difference in appearance, but it is easier to attach a homemade headboard to the wall. That also eliminates having to finish off the back of the headboard. If the bed is free-standing the headboard must be attached to the spring or bed frame.

Besides simplicity, the great thing about covering a headboard is that you control the amount of padding used. Brass bedsteads look terrific, but they are not comfortable for sitting up and reading in bed. The headboard you cover can be padded to any degree of softness. Use poly foam or batting.

A headboard sets the mood for the entire room. The traditional floral print used to pad it softens both the look and the feel of the wooden frame. The frame was designed with a padded insert, but the same effect can be had by cutting a pad from composition board, covering it, and attaching it to any plain headboard. *Courtesy Dan River*

The same idea is used here—a padded panel mounted on top of an existing headboard—but the look is different. The fabric design of sea and sand is right at home in a beach cottage. *Courtesy Burlington Mills*

Re-covering a Headboard

First strip off the old covering. Why not leave it on and cover over it? Because it will be difficult stapling over old staples or tacks, through multilayers of old coverings. If the old cover is plastic, the new fabric won't cling to it. This is an opportunity to refresh the old padding. It may not need to be discarded, but it no doubt needs a new top layer for buoyancy. Finally, it will give you an opportunity to see how the old covering was applied—and to see how simply a new one can go on.

This unattractive photo is included to show the skeleton of a headboard bought for a few dollars at a tag sale. Stripped of its shiny green plastic cover, it is nothing to be intimidated by. Beneath plastic was a thin layer of felt and under that nothing but the layer of corrugated cardboard seen in the photo. To simplify the lines, the curved top was sawed level. The bottom board, a later addition to the original, was knocked off.

Refreshed with a new covering of straw, the same headboard in its official "after" portrait. Re-covering a headboard is an easy project anyone can do.

Supplies and Tools Headboard, foam padding or poly batting or both if extra softness is wanted, fabric covering, scissors, staple gun and staples.

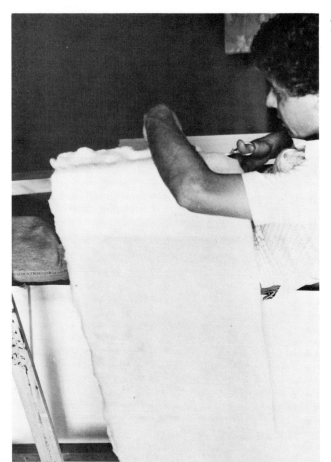

A layer of batting is stapled to the stripped frame.

125

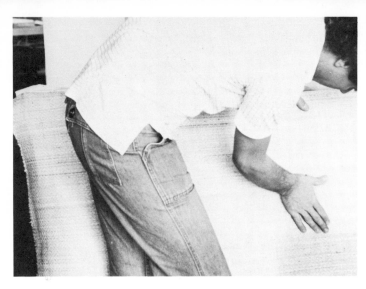

The new covering is straw. This was ordered from a fabric house; it comes by the yard and is available in different weaves and shades. Rug stores carry sisal intended for floor covering. The straw is held to the front of the headboard for general positioning. Because of the strong horizontal weave of the straw, care is taken to line it up exactly along the top edge.

Starting in the center of top edge, straw is pulled to back of frame and is stapled on. Corners are left loose.

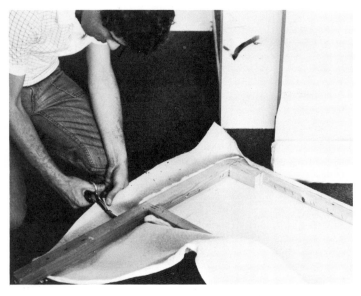

Extra straw is cut away at bottom. Bottom edge is stapled up next to point where leg posts begin. Scissors are used to clip straw into corner of leg.

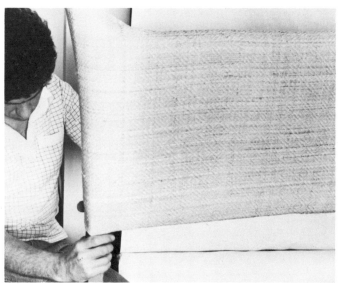

On front, hem is turned under at leg. Staples in back will keep the fold secure.

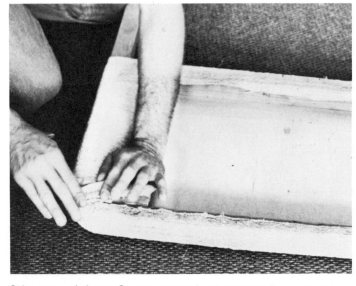

Sides are stapled next. Corners are eased and stretched. Straw is flexible, so tucks or pleats are not necessary.

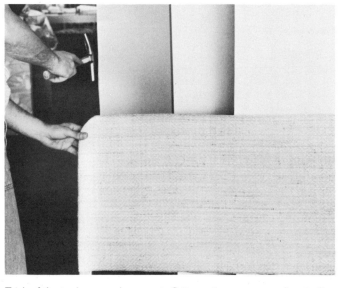

Trick of the trade uses a hammer to flatten out any unevens after stapling.

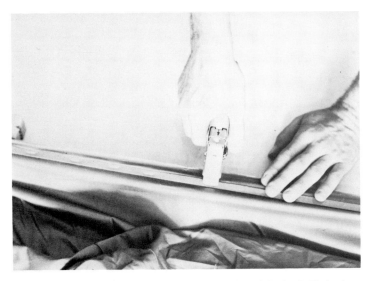

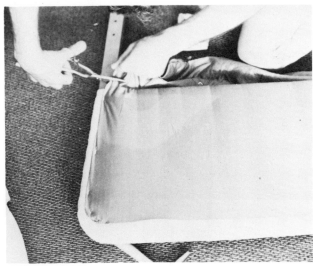

Cotton drapery lining is backtacked to the top edge of the back. Notice how wrinkled the lining material is.

The lining is turned to the right side, stretched tightly, and stapled flat along the sides and bottom, raw edges turned under. Tape or braid can be glued over the staples. The wrinkles in the lining have disappeared.

Making a Headboard

Making a headboard is as simple as requesting a piece of plywood cut to size at the lumberyard, taking it home, padding it, and stapling on a cover. A small electric saw can cut out any fancy shape that might be desired.

Padding and covering a plywood headboard of your own making is identical to the directions just given. To fasten it to a box spring or metal bed frame, bore holes in the plywood corresponding to the holes it is to line up with. If the headboard will be upholstered at these points, it is a good idea to drill the holes beforehand. After the padding and fabric are stapled in place, locate the holes. Use the point of a scissors to pry through the upholstery so screws or bolts can slip through easily.

This headboard went away to school. It was custom made with the special request that it be comfortable for reading in bed. The base is a panel of composition board covered with a layer of 1½-inch foam cut the same size as the panel. This was covered in one Indian print towel in bright earth tones. A fuzzy blanket used as a spread and a log for a bed table go well with it.

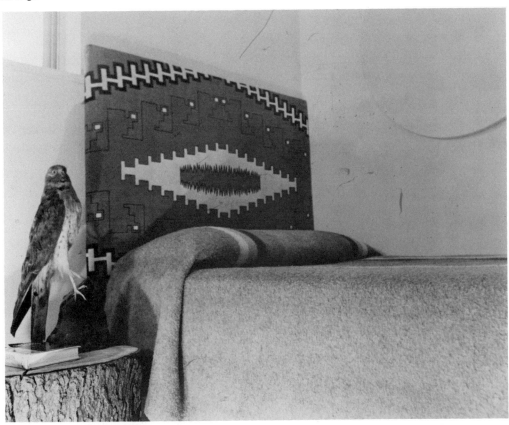

The desert sun motif is repeated in the shape of this circular headboard for a double bed. *Courtesy Burlington Mills*

Wall-Hung Headboards

Wall-hung headboards have advantages. The bed moves easily for making or cleaning around. Two nails in the wall will hold them up. Best of all, wall-hung headboards eliminate any outside limits on size. Make a headboard just large enough to rest a weary head upon or a headboard as large as a whole wall. Headboards that are to be fastened to the wall can be made of less sturdy materials than wood. Compressed layered cardboards or corrugated cartons can be used and padded with foam and batting. Since they will be attached to the wall, there is no need to finish the backs.

Canopies

Canopies over beds add an air of importance to a room. And symbolically they represent a comforting sense of shelter. The number of canopied or overhead framed beds in existence is small, but it is not difficult to create the illusion of a canopy.

This elaborate canopy wall features the built-in look. The quilted theme begins on the bench and is carried onto a headboard that extends to the ceiling. Quilting covers the front of a wood-framed canopy, which also serves as a baffle for reading lights. The room was planned by designer Cindi Mufson using Marimeko sheets. *Courtesy Celanese House*

A canopy can be formed by attaching 1×2 boards to the ceiling and stapling a hanging onto them. If the bed is in a corner, only two sides of the frame need to be put up, corresponding in length with the width and length of the bed. (For a free-standing bed attach a third piece of lumber so that the size of the bed is duplicated on the ceiling.)

No matter how light the hanging material is, don't depend on nailing or screwing the frame to the ceiling. If possible anchor into the rafters. Otherwise use enough toggle fasteners to guarantee that the canopy will stay up.

The look of a fourposter bed can be created by hangings at the four corners. A quick way to do it is to saw up an old picture frame into four corners. Screw wooden knobs or finials onto each cut end. Staple sheets to the corners and mount them on the ceiling. *Courtesy Utica Fine Arts sheets by J. P. Stevens*

With careful measuring and folding and with the use of fusible webs, the hangings can be made without the need for sewing. The hanging fabric is used double so that the view from inside the bed is as good as from outside.

Using the backtacking method in a horizontal way, attach the pressed, folded hanging to the outside of the frame. Staple into the wrong side of the top edge so that it will flop down over itself, hiding the staples. If a braid or self-welting trim is used, attach the hangings directly onto the outside of the frame, using the flat stapling method.

The Completely Upholstered Bed

This upholstered bed, tucked into the natural alcove formed by the eaves, is completely covered in white quilted cotton—a fitting base for an antique American quilt. Quilting is repeated at the windows in a more unusual way—chintz, loosely quilted on the diagonal, is used for window drapery. Although the room has many period touches, the look is definitely contemporary. The country home of Anthony Tortora and Jay Crawford. *Photo by Richard Champion*

A new bed covered totally in nubby white linen borrows its style from the traditional sleigh bed. *Courtesy Head Bed, Avery Boardman, Ltd.*

The simple classic lines of this upholstered bed take on an oriental feel when accessorized with a fan, a lacquered table, and a bamboo plant. *Courtesy Wamsutta*

This bed has simple, modern squared-off lines. A similar one can be made by constructing two simple plywood and frame boxes and bolting them to the head and foot of a box spring. Construct the boxes like the one in the directions for the module. Because they will hold the weight of the bed, use ¾-inch plywood framed with 1 x 4 lumber. After they are bolted on, face the open surfaces with ¼-inch plywood. Follow the directions for re-covering a headboard. *Courtesy Utica Fine Arts for J. P. Stevens*

In an apartment overlooking Gramercy Park, interior designer Harvey Herman beautifully combined an office and a bedroom. He covered the walls in a charcoal gray French menswear flannel. A plywood and planking platform was constructed and upholstered in gray tweed carpeting. Raising the platform a few inches gives it a floating look and also prevents stubbed toes.

Other Bed Bases

The water bed, the Japanese tatami mat, Scandinavian beds supported by flexible wood slats—new things to sleep on are crowding the standard box spring. The possibility that it might be healthful to sleep on a solid foundation brings new ideas of what a bed ought to look like. This trend sometimes lowers the bed—resting the box spring on the floor—or even eliminating it and putting the bed directly at floor level.

The Platform Bed

The bed on a platform can fill different roles—it can become a raised floor, it can give a bed a thronelike setting, or by substituting for bedstead or springs it can be a money saver. Most of all it gives a casual contemporary look to a room, especially when carpeted with the same carpeting as the floor—and perhaps the walls. A platform can be built from plywood and lumber. It is nothing more than an open box resting upside down on the floor.

133

7 Tables

Tables give off good feelings—dining tables heaped with holiday food, a bedside table just large enough to hold a lamp and one good book, a low table with precious mementoes and bibelots displayed as for a still-life painting. Even the common coffee table carries with it an association of sociability. But tables are first of all functional. Because they have no moving parts, no springs or cushions for comfort, and because their very use calls for a simple form, tables are the easiest kind of furniture to make.

A recent development sweeping the furniture field is the "nonwood" look table. This may have come from the dwindling supply of fine woods and fine cabinet craftsmanship, or perhaps because it's a way to add color and softness to our rooms. Whatever, tables covered in fabric, raffia, leather, are all around. This chapter shows how easy these tables—and others—are to make, to cover, to refurbish.

Any of the tables in this chapter can be padded or not. If padded, a glass top placed on top will give it protection. If unpadded, a hard sealer finish can be used.

Take the common cube. Over the last couple of decades it has been made in Plexiglas, in chrome, and in leather. It will be around for the next decades in as yet undreamed-of materials.

Opposite page: Stephen Chase, California interior designer, designed this room for a Texas family that loves color. A generously proportioned table sets the stage, covered in multicolor striped cotton. Glass protects the table top. *Courtesy Arthur Elrod Associates Incorporated. Photo by Max Eckert*

This cube, covered in batik cotton, can be copied very quickly. Made from five pieces of composition board, it is glued together and the box is then wrapped in a single piece of cotton fabric. The box's shape and size are determined by the fabric.

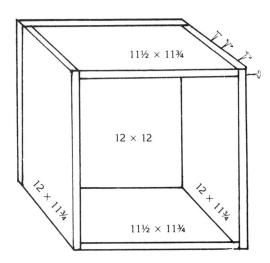

From ¼-inch composition board cut out these five pieces, measuring accurately: one 12 × 12 inch, two 12 × 11¾ inch, two 11½ × 11¾ inch. For sanity's sake, mark each piece with its measurements as you cut it. Fit them together as shown. Spread white glue along all joints, hammer in pins, and let set until dry.

Wooden Cubes

Unpainted furniture stores carry ready-made wooden cubes in a variety of sizes. Cubes can be made easily out of ¼-inch or ½-inch plywood following the method shown above to make the 12-inch composition board cube. Change the dimensions to suit—but remember to take into account the thickness of the plywood.

For a 24-inch cube made of ½-inch plywood needed are: one piece 24 × 24 inches, two pieces 24 × 23½ inches, and two pieces 23½ × 23 inches. If tools for cutting these pieces from a 4 × 8-foot sheet of plywood are not available, a lumberyard will make the cuts for a small fee.

Assemble the pieces in the same way that the cardboard pieces were put together. Glue the joints with white glue, but use small finishing nails instead of pins to reinforce. The nails (unlike the pins) stay, so hammer them in all the way.

With the right side of the cube facing up, center the fabric, using baste staples to hold it. Staple the fabric near the inside center point of each of the four sides. At the corner distribute the fabric so that it is evenly divided. A true bias will lie directly over the corner edge.

Fold the fabric and staple to the inside of the cube. Unless the fabric is bulky the excess will not need to be trimmed away. Turned over, the covered cube becomes a table on a table and can serve as a mini-museum for cherished souvenirs.

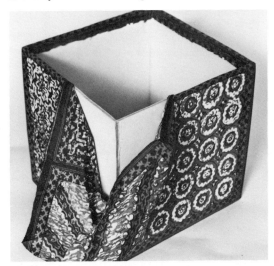

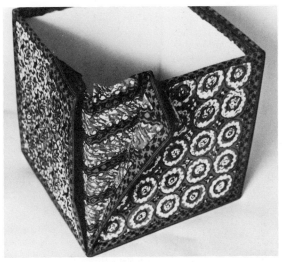

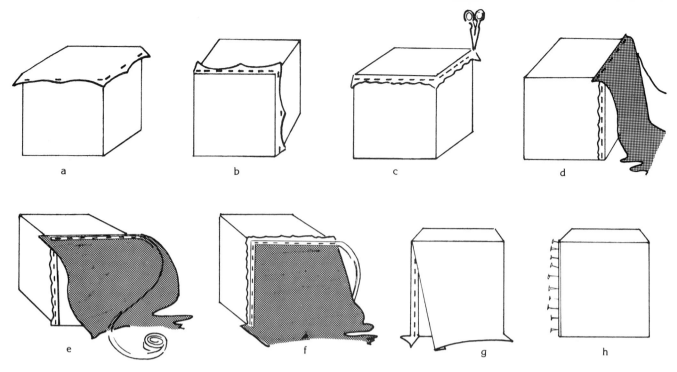

a   b   c   d

e   f   g   h

The neatest way to cover a cube uses only two pieces of fabric. a. Cover the top with a square of fabric one inch larger on all four sides. Flat staple it in the center of each side and at each corner. b. Turn cube on side and staple closely. Set staples in from the top edge about a half-inch. Staple all four sides this way. c. Clip excess fabric off at corners. d. Cut a piece of fabric large enough to wrap around the four sides, allowing an extra inch on each. Leave fabric extending one inch at the first corner and backtack the fabric wrong side up along the first edge. (Shaded area indicates wrong side of fabric.) e. Cover this row with upholsterer's tape with the tape edge level with the cube edge. f. Turn the cube and staple the second side the same way. Turn and staple the last two sides. g. Staple the one-inch extension where you began to the side of the cube. Trim away extra fabric from the top end so that the fold has a half-inch turnunder. h. Glue and pin the edge until dry. Turn cube over and staple along the bottom, mitering the corners.

This method uses three pieces to cover the cube. a. Cut a piece of fabric big enough to wrap around three sides with a one-inch allowance on all sides. Turn cube upside down and flat staple the fabric to the bottom of one side. b. Stretch the fabric over the top and around to the bottom. Staple this edge. On each side, fold the allowance under and staple. c. Cut a piece of fabric as large as the side plus an inch allowance for all sides. d. Backtack the fabric wrong side up to the top edge of the cube. e. Fold in allowance at each side and stretch the fabric tightly down to the bottom of the cube. Staple bottom. f. Part fabric along the sides, spread with white glue, and pin until dry. Do the opposite side the same way. With a scissors, trim the turned-back fabric away, leaving a half-inch allowance. Make a slash or two in the allowance almost, but not quite, to the corner so that it will lie flat when turned under. Following the crease line turn under the allowance. g. The finished cube.

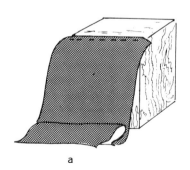 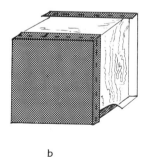 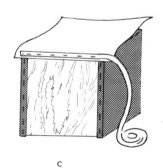 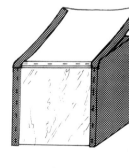

a   b   c   d

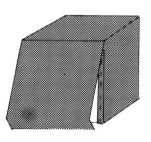 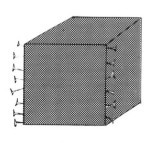 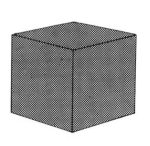

e   f   g

A house is not a home until it has a spot to keep the key for the camp trunk, the spare sink faucet washer, the mail . . . This table in an entry was made from a carpet tube. A circle was cut from composition board and glued on top. The tube was wrapped in a layer of batting. To duplicate, follow the same directions for covering a cube using the two-piece method.

A veritable strawberry patch with legs, this table was rescued from a junk dealer and given a new home.

Parsons Table

The Parsons table, named by the interior design school in New York City and not for the clergy, has been a mainstay in the furniture world for fifty years. It has a fine facility for adapting itself into any surrounding and its mood is controlled by the choice of covering.

It can be covered in two ways—using one piece of material wrapped and folded around it, or top and legs can be done separately. Both methods are easy. The one-piece method should not be used if the covering has a pattern that reads in one way. (If the roses grow in one direction they will be upside down on two of the legs.) Less fabric is needed for the pieced method.

This project used an old table 30 inches square and 16 inches high. (The next project uses a homemade one.) The chintz fabric chosen to cover it happened, at 37 inches, to be just wide enough to cover the top, the 3-inch side aprons, and to wrap onto the under surface.

Supplies and Tools 30-inch square table, 2 yards 37-inch-wide fabric, shade lamination, scissors, iron, staple gun and staples.

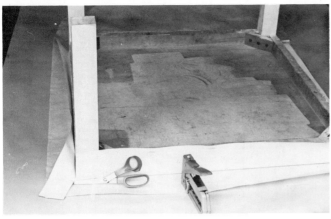

Because this table surface was uneven from dents, cracks, and paint blotches, the top surface was treated to a layer of the heat-sensitive laminate made for window shade making. Cut a piece of the laminate large enough to cover the table top and side apron, and wrap onto bottom edge. This photo shows the protective white paper that comes with the lamination left on at this stage to protect the glue side, which tends to be sticky. The laminate is used both to provide an ultrasmooth table surface and to keep the fabric from slipping or stretching. It is not necessary to use it. With the glue side facing work surface, center the table on it upside down. Staple the laminate to it. In this case do not work from opposite sides alternately, but do one side first. Start at the center of one side and staple out to the ends. Now stretch the lamination VERY tightly and staple it down on the opposite side. Do the other two sides.

To cover the legs, cut four pieces of fabric five times the leg width plus a half-inch allowance. (One leg surface gets a double layer.) To determine the length of the leg piece, measure the height of the table top and add 4-inch allowance. Position the fabric on one leg and staple it as shown in the photograph.

Wrap the fabric around the leg, turn under the hem allowance, and staple along the inside edge. With a scissors, clip the fabric at the inside top so that it fits snugly to the leg.

Trim away extra fabric around the table apron. Cover the other legs the same way. Cut the fabric for the top section large enough to cover the top and the apron, and wrap around to inside surface. The narrow width of this fabric required two sides to be stapled just to the bottom. If there is enough fabric, it should be brought around sides and bottom and to the inside. Place top piece of fabric upside down on work surface and center table on it upside down. (Remove lamination lining paper.) Starting from the center, staple the fabric on the same way that the laminate was done. Leave corners free.

Turn table right side up and fold back corners as shown in photo. With fingers, make a crease in fabric that runs exactly from top corner to the point where apron and leg form a right angle. Do this step with care, but without stretching the fabric. With a scissors, trim the turned-back fabric away, leaving a half-inch allowance. Make a slash or two in the allowance almost, but not quite, to the corner so that it will lie flat when turned under. Following the crease line turn under the allowance.

Turn table on side. Run a bead of white glue under the fold lines. With scissors, clip fabric so that it fits snugly to inside legs and apron bottom. Staple in place. Do the same steps at each corner. Wrap the extra fabric at the bottom of the legs as a package would be wrapped, mitering the corners. Glue in place and attach a furniture glide to the bottom. Or staple bottoms in place instead of glueing, and iron on a square of shade laminate, which will cover the entire bottom and prevent staples from scratching floor.

Another Way to Cover a Parsons Table

Supplies and Tools A 20×38-inch piece of ½-inch plywood, two pieces of 38-inch-long 1 × 3, two pieces of 18½-inch-long 1 × 3, four pieces of 15½-inch-long 3 × 3, 2¼ yards of 45-inch-wide fabric, glue, nails, hammer, staple gun and staples. Although the table is a simple design it requires some woodworking skill, which this book does not attempt to supply. If the skill and the tools are not available, a table can be bought at an unpainted furniture store.

Laura Ashley's sitting room in her flat in Chelsea, London. The parsons table again, this time in a different size and a different manner. Laura Ashley calls her small orderly designs "cottage prints." They appear to be colored with root and berry dyes—an influence of her Welsh country origin. The moss and flower fabric covering the walls is stretched on shutter panels at the ceiling-high windows. The directions follow for making a table exactly like the one in this photograph.

The photo shows the construction of the table.

Position fabric so that design is centered. Secure fabric with baste staples. A desk stapler opened up was used because of its small-size staples.

Turn table over and wrap fabric to insides, stapling randomly while adjusting. Trim away extra fabric as you work.

Fold corners into the angle formed by apron and leg as shown in the strawberry parsons table. The edge of the leg will be on the true bias of the fabric. Trim away extra fabric from leg. Use as many baste staples as necessary. Staple along inside corner of leg. Make clips to ease fabric as they are needed. Glue and pin the last seam along the inside of the leg. A piece of glass protects the table top.

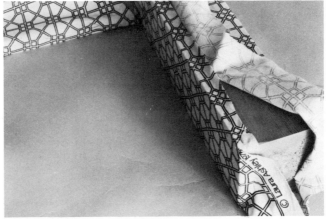

Imari Table

The table base is plywood. The fabric is wrapped carefully and accurately around the outside of the table and held in place with baste staples. Next it is stretched and stapled along the inside of the table. The inside lining is held in place with baste staples and then glued. The baste staples are removed when glue has set.

When Connecticut interior designer Ron Cacciola discovered this print fabric, it immediately set the focus for the theme of this airy porch. He designed the simple bold table to complement the fabric. The height, width, and length of the table is matched to the repeat of the fabric. Notice that a large medallion motif is centered on the end of the table and the border stripe runs exactly along the side edge. Once the table established the theme, Mr. Cacciola put together the other ingredients with a breezy kind of savoir faire. Paper lanterns and canvas blinds click beautifully with lavish greenery and Imari porcelain. *Photo by Bill Rothschild*

This table is covered in a peach-colored iris print fabric. The fabric is stapled onto a stock unpainted base. Glass on top would give it a good writing surface. *Interior designer, Ron Cacciola*

Picnic Table

This is a superfast way to dress up a dull picnic table for a special party. The effect is festive and it eliminates windblown tablecloths. The cloth is made from a sheet designed for Utica.

Turn the table upside down on the upside down fabric.

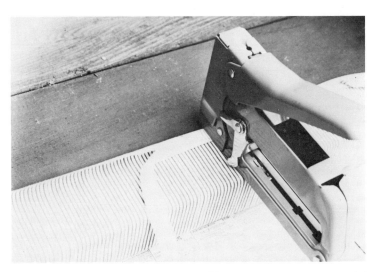

In order to be able to quickly strip the cloth off for washing, a cotton tape is stapled on at the same time. Any cloth tape or even the selvage cut from the sheet will work.

Carpet-Covered Tables

There's nothing new about carpet-covered tables. Early American housewives laid carpets on their dining tables as status symbols as fast as the clipper ships brought them from the Orient. Today's carpeted tables have a decidedly modern look. Flowing up in an unbroken line from the floor carpeting, they give the room a clean, uncluttered look.

Inevitably there are leftover pieces of carpet when a rug is installed. It is possible to cover the table with these pieces or to plan a table base to fit the remnant.

With a sharp utility knife, cut and fit the pieces to the table base. Spread rug adhesive on the table (or rug adhesive spray). Replace the carpeting and staple it securely. Spread the yarn so that the staples are shot directly through the carpet backing. The rug pile will cover staples.

Opposite page: This penthouse is designed by Yvette B. Gervey. The wool carpeting is called ''Jeans'' because it is the color of a pair of much loved, faded blue jeans. An octagonal base covered with carpeting is topped with thick glass—a good idea that can be copied so easily. The covering on the table and the built-in banquettes appears to be identical to the floor carpeting. Actually it is specially woven without a backing so that it may be used for just such purposes. *Courtesy Stark Carpet Corporation*

8 Windows

There are so many fresh and ingenious ways to treat windows that traditional draperies with triple pleats seem a tired choice. Not only is the initial investment high because of the large amounts of fabric required for these draperies, but cleaning costs for them are an ongoing expense. If the goal is to dress the window to be a center of interest or to make it recede through architectural trickery, if the intent is to frame a beautiful view or to hide a dismal one—there are a surprising number of ways to achieve any of these ends. This chapter deals with a variety of treatments, but all of them dispense with sewing and all use stapling.

Window with a View

The curtains have European-style drawstring tops. The hanging solution: gather curtains to correct size, staple them to a narrow strip of wood, and secure the board with a few small nails to the top of the window molding. The stapled side of the board faces the ceiling and the curtains fall gracefully over the frame.

Opposite page: These 8-feet-tall Swiss lace curtains cover windows overlooking a garden. The owners wanted to hang them for privacy only in the winter months when trees are leafless. Needed was a way to put them up and down seasonally without a hardware installation.

The same desire—for a garden view with privacy—was handled in a similar way, but the result is quite different. Casement panels of Belgian linen hang over the three windows in this studio living room. In this case they are stretched flat top and bottom for a tailored treatment. Designer, Emy Lesser. *Courtesy Belgian Linen Association*

Shades

Shades are an excellent choice for windows. Pulled down they provide privacy—pulled up they seem to evaporate. A recent study shows them to be an excellent insulator effective both against cold in winter and heat in summer. And they can be beautiful, too.

Using an iron-on shade lamination, it is possible to make colorful window shades out of any fabric imaginable. The Tontine brand lamination comes in two types—room darkening or translucent.

Locate a work surface at least as large as the size of the window opening. A hard-surface floor will do. If the surface is not heatproof, spread layers of opened newspapers over it. Spread out the shade laminate with glue side facing up. Place the fabric on it right side facing up. Adjust and smooth fabric until it lies perfectly flat.

Opposite page: David & Dash fabric is used in this room for a window shade made with lamination. Braid trims the bottom of the shade and borders the entire window. *Courtesy Window Shade Manufacturers Association*

The windows in this room have a sea view on either side: water outside the window, a shore-printed shade on the inside. The shades are made from sheets. *Courtesy Burlington Mills*

This berry patch wallhanging fabric used for both wall and shade makes a very dramatic statement in an otherwise dull box of a room. If shade is made carefully, it is possible for wall and window to be a continuous unbroken sweep of the design pattern. Most windows have a wooden frame around them, which gives more leeway for allowances to turn around roller at top and to form slat pocket at bottom. Directions for doing the wall are on pages 58 to 60. *Courtesy Materialize by N. Erlanger, Blumgart Company*

Note: Because it is cut from the wall mural fabric, this shade does not have the top and bottom allowance normally suggested for making a shade. When working with other fabrics where amount and placement are not restricted, cut the fabric 6 inches longer than the height of the window opening and one inch wider than the finished shade to allow for squaring and trimming.

Supplies and Tools Fabric as wide as the window and 6 inches longer, shade lamination material 18 inches longer than the height of the window and 2 inches wider, window shade roller at least 1⅛ inches in diameter, brackets, a 1¼-inch wood slat, chalk, yardstick, square, pencil, scissors, masking tape, iron, staple gun and staples. To determine the exact shade size, install brackets first and then cut roller to size. Press fabric, with a dry iron.

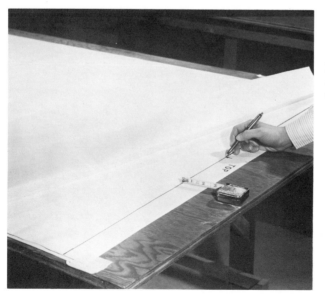

From laminate cut one 2-inch strip and one 8-inch strip for slat and roller attachments and put aside until later. Place laminate on work table with adhesive side up. Remove liner paper and save for pressing. Measure down 2 inches from the top and draw a line.

Place top edge of fabric on marked line, adjust and smooth fabric over the laminate until it lies perfectly flat. Set iron for the temperature suited to the fabric. Use scrap of fabric and laminate to test temperature. (If fabric weave does not allow enough heat to penetrate or if the temperature setting is too cool for proper bonding, turn shade over and, using lining paper, press the laminate side.) Start with low heat and begin to iron the fabric from the center out. The low heat will allow any bubbles or wrinkles to be smoothed out. After fabric has bonded lightly, place lining paper over fabric. Increase the temperature of the iron to the highest setting suitable. Now iron slowly with heavy hand over every inch of the shade. Allow to cool before handling further.

A shade that has not been accurately measured and trimmed to size may roll off center. With chalk, mark the finished width on the fabric side of the shade. Use a yardstick to draw a vertical line down each side of the shade to outline the width. Recheck top and bottom edges. Top edge should be squared off.

To finish the bottom edge, draw a line ⅓ inch from the bottom edge with chalk and fold along line so that laminate is against laminate. Anchor ends with masking tape. Using liner paper, press hem edge. Be careful not to press beyond the edge—the iron should not come in direct contact with either side of laminate. Remove liner paper from the 2-inch strip of laminate previously cut off. Place it along the folded edge of hem line, using chalk line as a guide. One inch should extend beyond each side edge and a small margin will extend above the hem edge. Place liner paper over laminate strip; press lightly. Insert wood slat, laying lower edge of slat flush with inside fold of hem. Replace liner paper; press firmly along upper edge of slat, giving special attention to the edges. Allow shade to cool.

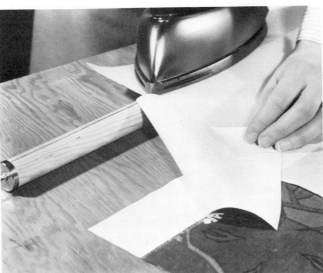

Remove the liner paper from the 8-inch strip of laminate. With adhesive side down, place lower edge of strip so it meets top edge of fabric. Cover with liner paper; press the 2 inches where bare laminate meets strip, holding upper edge of strip away from work table as you press. Allow shade to cool. Place upper edge of strip along the black line on shade roller. Staple laminate to shade. Cover with liner paper and press the laminate bonding it to the wood. Roll and press until the end of the adhesive is reached. Shade is ready to hang. *Courtesy Window Shade Manufacturers Association*

Handkerchief Drapery

The handkerchief drapery requires very little stapling but lots of ironing. The success of its uncomplicated construction depends on careful measuring and hanging. The handkerchief drapery is simply two panels of fabric with a jabot fold formed by catching the panels back from the window and fastening them to the side. The effect can be varied by fastening them high or low.

To make a handkerchief drapery, fold two panels of fabric in half lengthwise. The folded edge will be in the center of the window. The outside edges and the bottom hem are fused together with a ¼-inch strip of iron-on web. Since these draperies do not have the extravagance of large quantities of fabric, their success depends on proper hang. It is a good idea to interline them with a panel of fabric of substance, such as cotton flannel. Also, follow grain lines or the hangings will droop. The drapery shown in the photograph has a contrasting fabric on one side. In this case the center edge will not be a fold, but will have a hem formed with iron-on web the same as the outside edge. When the two panels are prepared, the unhemmed top is stapled to a narrow strip of wood cut to the width of the window. If a lining is used, be sure to catch this in when stapling. Hold the draperies to the window and roll the wood over until the panels are the proper length. Mount the wood strip to the top of the window frame, or on it, with small brads.

Cornices and Lambrequins

Cornices are frames used over windows to decorate or to hide the installation and hardware of curtains, draperies, or shades. Lambrequins are cornices that continue down the sides of the window.

A shaped cornice and informally pleated draperies are covered in a patterned fabric derived from the Egyptian Collection at the Metropolitan Museum of Art. *Courtesy Belgian Linen Association*

Opposite page: The handkerchief drapery is a classic style used in the past with formal fabrics. Today it is made from all sorts of fabrics and hung on windows of many proportions. This simple handkerchief hanging is made from two different cottons—a flower-printed front and a striped lining.

A lambrequin is customarily used over curtains or draperies. In this model room setting in Macy's Herald Square store, the curtains and the lambrequin are actually one and the same. Interior designer Joseph Widziewicz cut storybookish curtain shapes out of composition board and stapled striped cotton over them. Because it is a store interior, the windows and sunshine were created on the spot. The idea would work just as well used over real windows.

While they are usually constructed of wood, composition board and even corrugated cardboard may be used. A heavy appliance or TV carton, trimmed to size, padded with batting and with a covering stapled on, defies anyone to guess its origin.

The technique is explained in Chapter 3 in the section on covering pads. In this case the pad is shaped instead of being rectangular. Cut bases from composition board or thin plywood. Stapling is done on the back side of the panel. Staple the fabric along the long straight side first. The shaped sides are covered next, alternating between stapling and clipping with scissors where necessary to make the covering lie smoothly.

The heart of this home is its country kitchen. A window alcove is given all-stops-out treatment. Its side walls and banquette are upholstered in a provincial print. Plywood panels with window arches cut out are upholstered in the same print and outlined with braid. Before mounting the window panel, make draperies using iron-on fusible for bottom and side hems, then pleat and staple the drapery tops directly onto the back of the plywood. Floor is Solarian by Armstrong Cork Company.

Sometimes the architecture of a window makes it difficult to deal with. The problem here was not the window but the immediate scene outside. In order to let the light in without letting in the view, this arrangement was worked out—a blind with tiny perforations creates a starlight effect. A panel cut to fit the window recess is covered in fabric to suit the walls. The other window has double panels outlining and framing the blind. The same effect can be had by trimming with colorful cloth tape. *Courtesy Levolor Rivera Blinds*

Covered Rods

One way of dealing with window coverings involves hiding the hardware. Another approach includes the hardware as an integral part of the treatment. A good example of the "If you don't want to fight 'em, join 'em" school of decorating is the curtain rod. Currently interior decorators are emphasizing rods, not hiding them, by using fat, oversized ones. When the 1⅝-inch wooden closet pole size isn't big enough, plastic plumbing pipes are sometimes used.

Additional window treatments are described in other sections of the book. The fabric-hinged screen, although explained in the chapter on screens, is used at the window. Stretcher bars make perfect window shutters. They can be completely upholstered or they can have a simple panel of fabric stretched in the center. Sheets, already hemmed both top and bottom, can be stapled onto wooden rods using pleats, tucks, or gathers.

To cover a wooden rod, cut fabric one inch longer than rod and one inch wider than its circumference. Fold under half inch on both ends and on one long side. Press folds. With staple gun, fasten unpressed raw edge to the pole in a straight line. Roll fabric around rod and staple in place over raw edge. Arrange drapery material into pleats and baste them in place with a desk-type stapler. With staple gun, fasten pleated draperies to rod. Rotate rod until pleats roll under securely and staples are concealed.

This cabin in the woods got a burst of charm with instant curtains. A cardboard tube, donated by a fabric store, is covered with calico. Curtains are attached as shown in the previous drawings. Round knob on end of pole is a tennis ball wrapped in a circle of fabric, then glued into tube opening. A styrofoam ball used for making ornaments works, too. The whole treatment rests on an L-shaped bracket attached to the wall.

The same method can be used in a more traditional way. Wooden finials are screwed into wood poles. Using a staple gun for installation works well when the window hangings are stationary.

Opposite page: These dotted draperies were hung informally on wooden rods. Covering the rods with the same fabric adds a definite dash of style. *Courtesy of the Belgian Linen Association*

This vinyl floor was installed by a simplified process that involves stapling the flooring down at the edges of the room. The names of two hard-surface continuous floorings that can be installed this way are Premier Sundial and Tredway. *Courtesy Armstrong Cork Company*

9 Floors

Having reached the final chapter in the book, the reader will not be surprised to learn that a staple gun can be used to install floor coverings, whether they are hard surface, carpeting, or fabric.

Hard Surface

It is now possible to buy seamless hard-surface floor covering that can be installed by a do-it-yourselfer and a heavy-duty staple gun.

Move all furniture or appliances out of the room. Remove existing molding from around the floor. Unroll the flooring in an area where it can lie flat. The factory-cut edge of the flooring will be butted against the longest wall in the room. With chalk, transfer accurate measurements to the flooring material. Cut, adding at least a 3-inch allowance to all dimensions except the factory edge. Position the material in the room. Use a metal straightedge or carpenter's square to guide a utility knife.

Staples are spaced at 3-inch intervals. A heavy-duty staple gun is used here. The staples are concealed when the molding is replaced.

The finished room. The manufacturer says the flooring "forgives" an amateur's mistakes—up to a point. If slightly undercut, the material can be stretched. If cut too loosely, bulges or wrinkles will disappear as the flooring contracts. The flooring has an undercushion, so it can be installed over slight irregularities. *Courtesy Armstrong Cork Company*

Carpeting

Carpeting is used so extensively in interior furnishing that the term wall-to-wall may soon be abandoned in favor of **wall-to-ceiling.** Carpeting is used to cover platforms, tables, walls, and ceilings, and to line conversation pits. Two of the qualities that have made it popular are the texture it brings to a surface and its sound-deadening ability. This popularity is surprising because it is far more expensive to carpet a wall than to paint it. Manufacturers have responded by making a carpeting specifically for "upholstery" use. Since it does not need the wearability of regular floor covering, this type has the same appearance but a lighter backing and weight. Depending on the type and construction of the carpeting it is attached with a staple gun, sometimes in combination with rug adhesive.

Strictly for drama, interior designer Richard Plouffe added a floating platform around two sides of this living room. A corner island juts out to hold a wing chair and a brass desk. *Photo by Frederick E. Paton*

A young man's solution to a small one-room apartment: Go East. He borrowed from the Japanese the no-furniture look and the tatami mat. Instead of sleeping on a straw mat he used sisal carpeting wall to wall. When company comes, the mattress is covered with a throw and the platform surrounding the bed becomes seating, table, and lounge. *Courtesy Dan River*

Carpeting is used extensively in the living room of this Connecticut home. A raised platform in front of a large glass area becomes a greenhouse. Placed on the diagonal, a massive carpeted lounge is piled with cushions. The lounge was built by a carpenter to designer Richard Plouffe's plans, then upholstered by stapling on carpeting.

Carpeted platforms need not be low. This room in the Woolworth estate on Long Island was designed by Harvey Herman to be a bed/sitting room. At window height a platform holds a mattress covered in suede cloth. The view from the bed is of the original tracery ceiling.

Interior designer Rob Hardy accepted the challenge offered by a bare studio apartment—his own. Working from one small budget, the memory of one woodworking shop class in junior high, and lots of ideas, he met the challenge successfully. Planning was carefully done; measuring was accurately done. He was surprised at how easy it went along. The bed sits on a low platform. A higher one along the window wall serves as shelf and seating.

The other side of the same room shows that the window platform continued around the walls becomes the sofa. The same gray shag carpeting is used throughout.

Installing Carpeting

Ambitious do-it-yourselfers can install wall-to-wall carpeting using a staple gun (heavy-duty) instead of the customary tack strips. A knee-kicker is necessary to get the carpet stretched tightly, but it can be had from tool rental services for a small fee. Start in a corner of the room working out on either side, stretching as you go.

Because few of today's carpetings ravel or require finishing on the edges, an interesting mosaic rug can be created using a collection of sample squares or shapes cut from rug remnants. First work a pattern out to scale on paper. Then start from the center of the room, butting the edges and stapling them in place. Spread the pile away so that the staple is shot directly into the backing.

A custom-designed rug with two darker broad bands set into the area between the bunks is used in this boys' room designed by Arthur Leaman. *Courtesy Celanese Fortrel*

Covering Floors with Fabric

The development of clear finishes such as the polyurethane varnishes has made it possible to cover floors in fabric. While they are not as durable as other hard-surface floorings, fabric-covered floors are a possibility to consider for their decorative appearance and for their economy. It is recommended that they be given five coats of sealer, but even so they are not practical for high-traffic areas.

Not all fabrics are suitable as floor coverings. The main requirement is that it have a sufficient content of cotton or other natural fiber such as linen or burlap, which will allow the sealer to soak into the fabric. The fabric chosen should be a patterned one. Even careful preparation of the floor cannot eliminate planking or joints from showing through. Roofing paper or plywood can be used between the floor and the fabric if the effort and the expense seem worthwhile.

Often the reason for using fabric on the floor is to match it to the fabric on walls. Unfortunately there is no sealer that will not change the color. The sealers give a yellow or beige cast to the fabric. If the fabric is an earth-toned one, this will not make a difference, but if the colors are in the blue or pink family the change can be so drastic as to cause the walls and floor to clash. The sealer, while darkening colors, tends also to intensify and give a boldness to some shades. A test run with one coat of sealer will give a good idea of the result. The succeeding coats will not have much effect on the color.

Sheets are a good choice because of their large size. The fabric panels are not seamed together before being applied, as the seam allowances would form a ridge on the surface. Selvage edges must be cut off so that the pattern can be lined up perfectly. Cut carefully with a sharp scissors to prevent raveling. Once the sealer is on the edges will not fray.

Canvas, an old-fashioned floor covering, is back on the scene again. The charm of canvas is that it takes well to stenciling or other painted designs. This can be done either before or after the canvas is glued to the floor.

If the fabric has a strong pattern motif, the first panel should be centered in the middle of the floor. If the pattern is an allover one, it may be easier to start along one long wall.

Wax must be removed from the floor, which should be dustfree. Remove moldings, if possible. Apply white glue diluted with water to the floor on the area where the first panel will be. An extra pair of hands will make the job much easier. Stretching the panel as much as possible, lay it over the glue. Baste staple the edges and leave them in place until the glue is dry. There must be no wrinkles or bubbles in the fabric. Apply the next panel the same way, matching the pattern. Allow the floor to dry at least overnight before applying the first coat of sealer.

Opposite page: Since a king-size sheet covers an area 9 feet by 9 feet 2 inches the floors of most rooms can be covered by using one or two sheets. This elimination of seaming and pattern matching makes the job go fast. Perseverance—in applying enough coats of sealer to make the floor durable—is the hardest part of the job.

Sheets do not match up patternwise along selvage edges as do decorative yardgoods. Usually just a few inches must be trimmed away to match up pattern.

Spread the diluted glue (about half-water, half-glue) over a large area, then stretch fabric over it, stapling at edges so that it stays taut and smooth.

Index

STAPLE IT!
Easy Do-It Decorating Guide

by IRIS IHDE FREY

A new kind of decorating book, which reveals for use in your own home the techniques and tricks used by professionals to create eye-catching, crowd-stopping model rooms. With one simple tool—the staple gun—the author shows how you can duplicate the spectacular results of display artists and scenic and interior designers. Instructions accompanied by clear illustrations and how-to photos enable the amateur to decorate an entire house with professional results. Since the methods shown eliminate the need for sewing and a sewing machine, the book has value for a wide audience.

Here are techniques for using staplers to cover walls and ceilings with fabric and to make wall hangings, draperies, shades, and supergraphic panels. Included are dozens of ways sheets can be used for inexpensive but dramatic effect. This book shows how to transform furniture discards into treasures and how to turn ordinary crates into seating modules. You learn how to do your own upholstering, starting with the simple job of covering a chair seat and progressing to more challenging projects such as the total upholstering of a wooden chair. Several ways to treat parsons tables and cubes are shown. Included are imaginative window coverings that are both attractive and energy saving. For accessories, the author demonstrates how to staple boxes, picture frames, and screens. For screens, learn how to make rare "old world" fabric hinges. From ceiling to floor, from window to bed—every aspect of a well-dressed home is covered.

Along with the how-to photos are more than 100 other photos in both color and black and

(continued on back flap)